A Laboratory Manual

on

Virtual Experimentation on Electrical AC Machines and Circuit Networks using MATLAB/Simulink and MULTISIM

A Laboratory Manual

on

Virtual Experimentation on Electrical AC Machines and Circuit Networks using MATLAB/Simulink and MULTISIM

Arif Iqbal
Department of Electrical Engineering
Rajkiya Engineering College Ambedkar Nagar
U.P., India

G. K. Singh
Department of Electrical Engineering
Indian Institute of Technology Roorkee
Uttarakhand, India

CWP
Central West Publishing

Disclaimer
Every effort has been made by the publisher, editors and authors while preparing this book, however, no warranties are made regarding the accuracy and completeness of the content. The publisher, editors and authors disclaim without any limitation all warranties as well as any implied warranties about sales, along with fitness of the content for a particular purpose. Citation of any website and other information sources does not mean any endorsement from the publisher, editors and authors. For ascertaining the suitability of the contents contained herein for a particular lab or commercial use, consultation with the subject expert is needed. In addition, while using the information and methods contained herein, the practitioners and researchers need to be mindful for their own safety, along with the safety of others, including the professional parties and premises for whom they have professional responsibility. To the fullest extent of law, the publisher, editors and authors are not liable in all circumstances (special, incidental, and consequential) for any injury and/or damage to persons and property, along with any potential loss of profit and other commercial damages due to the use of any methods, products, guidelines, procedures contained in the material herein.

A catalogue record for this book is available from the National Library of Australia

ISBN (print): 978-1-922617-41-5

Preface

Electrical machinery and circuit network are taught as compulsory courses at undergraduate with their applications at postgraduate level in most of the technical institutions. Experiments based on the courses taught is an important aspect of practical implementation of working concept taught during class hours. This book deals with the most essential experiments related to important topics of electrical machines and networks which are practically performed by students in majority of technical institutes. Theoretical background of every experiment is reviewed before the system simulation. Detailed step by step experimental procedures are explained with necessary diagram. Development of experimental setup by using MATLAB/Simulink and MULTISIM has been explained from scratch, which also enhances the simulation and analytical skills for various practical systems. Complete content of book is presented in two parts: Part A and B.

Part A: This part of the book presents the experiments based on electrical machines i.e. induction motor (single and three-phase), synchronous generator and transformer. In this part, MATLAB/Simulink model has been developed and operated as per the objective of experiments. All the necessary procedural steps are explained and implemented for every experiment.

Part B: This part of the book presents experiment based on electrical circuit network i.e. network theorems and circuit response. In this part, MULTISIM is used to design the circuit model and operated as per the objective of experiments. All the necessary procedural steps are explained and implemented in every experiment.

Every step of simulation explained in both parts can be further extended to investigate and analyse for Academics, Industrial, and Research & Development purposes. This book can be used as a reference to simulate and analyse the virtually developed system using AC machines and circuit network by various level of students and working professionals.

Authors express a profound gratitude to our colleagues from both the institutes (Rajkiya Engineering College, Ambedkar Nagar and Indian

Institute of Technology, Roorkee) for their fruitful and valuable suggestions. Authors acknowledge their family members for moral support during writing process of the book.

Arif Iqbal
G. K. Singh

About the Authors

Arif Iqbal received his B.Tech. and M.Tech. degrees, both in Electrical Engineering, from Aligarh Muslim University, Aligarh, India, in 2005 and 2007, respectively. He has completed his Ph.D. from Indian Institute of Technology, Roorkee, India, in 2015. He is having an experience of eight years in industry and teaching in the field AC drives and power system. Currently, he is working as Assistant Professor in the Electrical Engineering Department, Rajkiya Engineering College Ambedkar Nagar, Akbarpur, India. His area of interest is multiphase AC machine & drives in general and high phase order ac machines and drives in particular, power electronics and renewable energy.

G. K. Singh received the B.Tech. degree in electrical engineering from G. B. Pant University of Agriculture and Technology, Pantnagar, India, in 1981, and the Ph.D. degree in electrical engineering from Banaras Hindu University, Varanasi, India, in 1991. He worked in industry for nearly five and a half years. In 1991, he became Lecturer at M. N. R. Engineering College (presently, Motilal Nehru National Institute of Technology), Allahabad, India. In 1996, he moved to the University of Roorkee, Roorkee, India. He was Visiting Associate Professor with the Department of Electrical Engineering, Pohang University of Science and Technology (POSTECH), Pohang, South Korea, and Visiting Professor with the Department of Electrical and Electronics Engineering, Middle East Technical University, Ankara, Turkey. He is currently senior Professor with the Department of Electrical Engineering, Indian Institute of Technology Roorkee, Roorkee-247667, Uttarakhand, India. He has been involved in the design and analysis of electrical machines in general and high phase order ac machines in particular as well as power system harmonics and power quality. Recently, a book entitled Computational Intelligence and Biomedical Signal Processing, co-authored by him has also been published by Springer Nature. He has coordinated a number of research projects sponsored by the Council of Scientific and Industrial Research (CSIR) and University Grants Commission (UGC), Government of India. Prof. Singh received the Pt. Madan Mohan Malaviya Memorial Medal and the Certificate of Merit Award at the Institution of Engineers, India, from 2001 to 2002. He secured rank 1 in India and 250 world ranking (top 0.15%) in the subject area "Networking and Telecommunications" as per the independent study done and published by Stanford University, USA, in 2020. In its' 2021 report on worldwide researchers, Stanford University, USA has placed him at 266 world ranking (top 0.11%) in the subject area "Energy".

Table of Contents

Part A

Experiments with MATLAB/Simulink

GENERAL INTRODUCTION OF AC MACHINES AND TRANSFORMER

AC motors and transformer are the important topics, which are dealt in the courses of electrical machinery taught at graduate level. These topics are also frequently considered at post graduate level and research work during the investigation of electric drive and power system. Before going through experiments on AC machines and transformer, it seems necessary to review the basic details related to their construction and operation which are briefly presented in following sections. Complete details may be found in literatures given in references.

A. THREE-PHASE INDUCTION MOTOR

Induction motor is an important member of AC machine, which is preferably used in many industrial and domestic applications. It offers many potential advantages, when compared with other types of motors, such as robust construction, low maintenance, lower cost and higher overload capacity.

A.1 CONSTRUCTION

Induction motor is constructed by two essential parts: i) stationary part known as **stator**, ii) rotating part known as **rotor**. Stator is made by using alloy steel (high grade) lamination to reduce eddy current loss. Inner periphery of stator is slotted, in which enamelled conductor is placed and are connected to form the three-phase armature winding. Three-phase winding may be in star or delta connected. Rotor on other hand is made by thin lamination of same material as stator. Laminated rotor core is mounted on the shaft and are slotted on outer periphery. Depending on the rotor construction, motor is classified in two categories: i) Squirrel-cage rotor or simply cage rotor type, ii) wound rotor or slip ring rotor type. In cage type motor, laminated iron core consists of slots to place the skewed conductors of copper or aluminum. Rotor conductor/bars are shorted through end rings to form a shape of cage. Skewing of rotor conductor offers the following merits:

    i)   Developed torque is uniform with reduced noise.
    ii)   Magnetic interlocking between the teeth of stator and rotor is avoided.

Most of the induction motors used in various industries are squirrel cage rotor induction motor.

Slip ring type induction motor consists of rotor, which is equipped with distributed three-phase winding connected in star on outer periphery. End terminals of rotor winding are connected to slip ring mounted on the shaft. Hence, rotor circuit resistance may be varied by connecting external resistance through brush. This type of motor is used, where high staring torque is required such as in refrigeration plants. Also, the speed and torque of motor can be controlled by inserting external resistance in the rotor circuit. Slip ring induction machine is also used as generator for small power production utilizing renewable energy sources.

A.2 OPERATION

Induction motor is operated, when a balanced three-phase input supply is fed to the armature winding (i.e. three-phase stator winding) resulting in rotating magnetic field in the airgap with synchronous speed. Rotating field cuts the rotor conductor and an emf is induced. Since, rotor conductor is stationary and shorted with end rings, hence current will start to flow through it. Therefore, current carrying rotor conductor placed in rotating magnetic field develops a mechanical force. Sum of developed force in all rotor conductors will produce a torque in rotor, and hence motor starts running. It may be noted that the direction of rotor rotation is same as that of airgap field and is governed by Lenz's law.

B. SINGLE-PHASE INDUCTION MOTOR

It is the most commonly used motor in domestic appliances like washing machine, fans, vacuum cleaner, mixers etc. Such motors have a lower power rating, typically one kilowatt or less than one horse-power, hence known as *fractional horse-power* or *fractional kilowatt motors.*

B.1 CONSTRUCTION AND OPERATION

Single-phase induction motor is very similar to three-phase machine. Unlike to three-phase, stator consists of armature winding, which is single-phase. Stator core is laminated to reduce the eddy current loss. Inner periphery is slotted to place armature conductor for winding configuration (main winding). Rotor on other hand is cage type, similar to three-phase induction motor.

Operation of single-phase induction motor can be explained by using Double-revolving field theory. It states that a stationary pulsating magnetic field may be resolved in two components of equal magnitude and rotate in opposite direction with respect to each other (say, forward and backward flux component). Magnitude of each field component is half of the stationary pulsating magnetic field. During starting process, torque produced by interaction with forward and backward flux component is equal and opposite, cancelling each other. Hence, net torque produced is zero and motor is not a self-starting. Motor may be started with rotating magnetic field in airgap which is obtained as an interaction with two alternating field fluxes with some phase angle difference. Hence, additional alternating flux is initiated by another winding known as auxiliary winding used during staring process. After picking up the sufficient rotor speed, additional alternating flux may be removed and motor will continue to operate under the influence of main field flux in the airgap. Single phase induction motors are classified according to the starting method used with auxiliary winding:

i)   Split-phase motor
ii)   Capacitor-start motor
iii)   Capacitor-start capacitor-run motor
iv)   Permanent-split capacitor motor
v)   Shaded-pole motor

C. SYNCHRONOUS MACHINE

Synchronous Machine rotates at a constant speed, dependent on supply frequency and number of poles. It is preferably used as generator and known as alternator.

Stator of synchronous machine is similar to that of three-phase induction machine, consisting of stationary armature winding. Stator core is laminated to reduce the eddy current loss, which consist of slots in inner periphery for three-phase winding configuration. Rotor construction is different and may be following two types:

i)   Salient pole rotor
ii)   Cylindrical rotor

Salient pole rotor consists of poles, which are projected from the surface of rotor core. Rotor is made by steel laminations to reduce the eddy current losses. Rotor field coils are placed across each pole. Each coil is connected in series to constitute rotor winding. Synchronous machine with salient pole rotor has non-uniform airgap. Under pole center, airgap is minimum, but it is maximum in between the poles. Salient pole rotor has large diameter and smaller axial length. Machine with salient pole rotor has higher number of poles and operates at lower speed.

Cylindrical rotor machine is also known as non-salient pole rotor machine. Such rotor is constructed in the form of smooth cylindrical shape and has uniform airgap. Rotor is made from nickel-chrome-molybdenum (high grade). About two-third of rotor periphery is slotted at regular interval to rotor conductor for field winding (distributed type). Unslotted portion of outer rotor periphery constitute the rotor pole (two or four poles). Cylindrical pole rotor has small diameter and larger axial length. Cylindrical rotor is used in high speed machine.

During the operation of synchronous generator, rotor is driven by prime mover at synchronous speed using a turbine (stream or gas). DC field excitation is provided to rotor winding through slip-ring brush arrangement. Hence, field flux will cut the armature conductors and alternating three-phase voltage is generated in the stator windings.

D. TRANSFORMER

Transformer is a static device, which consists of two (or more) electric circuits interlinked by a common magnetic circuit to transfer the electrical energy between them. Basically, a simple transformer consists of two types of winding: Primary and Secondary. Input voltage is fed to the primary winding, which establishes the magnetic field flux due to current flow. Magnetic flux links to the winding of secondary, and an emf is induced. Hence, the electrical energy is transferred from primary to secondary winding through mutual coupling. Voltage induced in secondary may be more or less than primary, which depends on number of turns of secondary with respect to primary. If number of turns of secondary is more than primary, transformer is known as **Step-up** transformer and, if number of turns of secondary is less than primary, transformer is known as **Step-down** transformer. It may be noted that there is no relative motion between primary and secondary; and therefore, frequency of induced secondary voltage is equal to primary input supply.

EXPERIMENT 1

1.1 Objective: To perform no-load, blocked rotor and DC test on three-phase induction motor.

1.2 Software Required: MATLAB /Simulink software.

1.3 Brief Theory

Determination of motor efficiency requires a direct operation at different loads, but it results in power loss (particularly in larger motors). Hence, indirect tests are suggested to determine the efficiency of three-phase induction motor: No-load test and Blocked rotor test. These tests are based on the evaluation of circuit parameters of equivalent circuit of three-phase induction motor on per phase basis, shown in Fig. 1.

1.3.1 No-load Test

This test is similar to no-load test of transformer, where three-phase induction motor is allowed to run at rated voltage under no-load. Input line voltage and current is measured by connected voltmeter and ammeter respectively. Input power is measured by using two wattmeter method. Circuit connection of no-load test is shown in Fig. 2.

During this test, following observation is noted

Voltmeter reading = V_{nl}

Input phase voltage $V_{np} = V_{nl}/\sqrt{3}$ (1.1)

Ammeter reading = I_0 (i.e. no-load current)

Power input $P_{nl} = P_1 + P_2$ (1.2)

Therefore, power factor at no-load is given by

$$\cos \Phi_0 = {P_{nl}} \Big/ {\left(\sqrt{3} V_{nl} I_0\right)}$$ (1.3)

Active component of no-load current $I_a = I_0 \cos \Phi_0$ (1.4)

Reactive component of no-load current $I_r = I_0 \sin \Phi_0$ (1.5)

Hence, parameter of no-load branch parameters are given as

$R_w = V_{np}/I_a$ (1.6)

$X_m = V_{np}/I_r$ (1.7)

1.3.2 Blocked Rotor Test

This test is similar to short-circuit test of transformer, where rotor is locked externally and input terminals are fed by a reduced voltage and frequency to allow the current flow in motor winding at rated value. In this condition, rotor winding is shorted for wound rotor, which is permanently shorted for cage rotor. Circuit connection of blocked rotor test is shown in Fig. 3.

During this test, following observation is noted

Voltmeter reading = V_{sc}

Ammeter reading = I_{sc} (i.e. no-load current)

Power input $P_{sc} = P_1 + P_2 = \sqrt{3}V_{sc}I_{sc} \cos \Phi_{sc}$ (1.8)

where, $\cos \Phi_{sc}$ is power-factor at blocked rotor (i.e. short-circuit)

Hence, equivalent resistance referred to stator per phase is

$R_{e1} = \dfrac{P_{sc}}{I_{sc}^2}$ (1.9)

And equivalent impedance referred to stator per phase is

$Z_{e1} = \dfrac{V_{sc}}{I_{sc}}$ (1.10)

Equivalent reactance referred to stator per phase calculated as

$$X_{e1} = \sqrt{Z_{e1}^2 - R_{e1}^2} \qquad (1.11)$$

1.3.3 DC Test

In addition to above tests, a DC test is also conducted to determine the stator winding resistance R_1 per phase. For this purpose, DC supply is fed to end terminals to any two phase windings say *a-b*. Complete circuit representation of test is shown in Fig. 4. In the circuit, a resistive element in the form of lamp is connected in series to limit the current flowing into the winding.

During this test, following observation is noted

Voltmeter reading = V_{dc}

Ammeter reading = I_{dc}

Winding resistance per phase is given by,

$$R_{dc} = 0.5 \frac{V_{dc}}{I_{dc}} \qquad (1.12)$$

Obtained DC resistance R₁ is, then corrected for 'Skin effect' to get AC resistance:

$$R_1 = (1.01\text{-}1.15)\, R_{dc} \qquad (1.13)$$

This correction is only applicable to squirrel cage rotor motor only.

1.4 Experimental Procedure using MATLAB /SIMULINK

In this experiment, a 3 HP, 220 V, three-phase induction motor is selected having the parameters given in Table I.

1.4.1 No-load Test

Step 1: Launch the Simulink in Matlab.
Step 2: In new model file, add the required blocks of 'Asynchronous Machine' together with other measuring blocks, which are to connected as shown in Fig. 5. Motor parameters are set into machine model through 'Block Parameters' dialog box as shown in Fig. 6.

Step 3: Run the developed test model at rated voltage under no-load condition for some 2 sec.

Step 4: Take the reading of input voltage, current and power, as shown in Table II.

1.4.2 Blocked Rotor Test

Step 1: Simulink model used for no-load test is also used in blocked rotor test as shown in Fig. 7. In the model, rotor is locked by setting the infinite value to moment of inertia.

Step 2: Adjust the input voltage at reduced magnitude for rated current in motor windings.

Step 3: Take the reading of input voltage, current and power, as shown in Table 2.

1.4.3 DC Test

Step 1: Between the any two end terminals say *a-b*, connect a DC voltage source (100 V in this case), a resistive element (in place of lamp of 10 Ω) along with other measuring units shown in Fig. 8. In the circuit, a resistive element is also added to complete the circuit.

Step 2: Take the reading of input terminal voltage and current, as shown in Table 2.

Step 3: Calculate the value of winding resistance per phase.

1.5 Sample Calculation

Results obtained from tests of three-phase induction motor using mathematical steps are:

1.5.1 From DC test

$$R_{dc} = 0.5 \frac{V_{dc}}{I_{dc}} = 0.5 * \frac{7.988}{9.201} = 0.434 \, \Omega$$

$$R_1 = 1.01 * 0.434 = 0.438 \, \Omega$$

1.5.2 From No-load test

Power factor at no-load is given by

$$\cos \Phi_0 = P_{nl} \Big/ \left(\sqrt{3} V_{nl} I_0\right) = \frac{41.94}{\sqrt{3} * 220 * 5.669} = 0.0194$$

$$I_a = I_0 \cos \Phi_0 = 5.669 * 0.0194 = 0.110 \, A$$

$I_r = I_0 \sin \Phi_0 = 5.669 * 0.9998 = 5.6679 \, A$

$R_w = V_{np} \, (per \ phase)/I_a = \dfrac{220/\sqrt{3}}{0.110} = 3464.1 \, \Omega$

$X_m = \dfrac{V_{np}}{I_r} = \dfrac{220/\sqrt{3}}{5.6679} = 67.2297 \, \Omega$

During no-load test, per-phase equivalent circuit and phasor diagram of motor are shown in Fig. 9 (a) and Fig. 9 (b) respectively.

1.5.3 From Blocked rotor test
Power factor is given by

$\cos \Phi_{sc} = {P_{sc}} \Big/ {(\sqrt{3} V_{sc} I_{sc})} = \dfrac{326.10}{\sqrt{3}*28.77*9.50} = 0.6889$

$Z_{e1} = \dfrac{V_{sc} \, (per \ phase)}{I_{sc}} = \dfrac{28.77/\sqrt{3}}{9.50} = 1.7485 \, \Omega$

$R_{e1} = Z_{e1} \cos \Phi_{sc} = 1.7485 * 0.6889 = 1.2045 \, \Omega$

$R_2 = R_{e1} - R_1 = 1.2045 - 0.4383 = 0.7662 \, \Omega$

Total leakage reactance,

$X_{e1} = Z_{e1} \sin \Phi_{sc} = 1.7485 * 0.7249 = 1.2675 \, \Omega$

Considering,

$X_1 = X_2 = \dfrac{X_{e1}}{2} = \dfrac{1.2675}{2} = 0.6338 \, \Omega$

During blocked rotor test, per-phase equivalent circuit and phasor diagram of motor are shown in Fig. 10 (a) and Fig. 10 (b) respectively.

1.6 Results and Discussion

From the calculation of the experimental readings, results are given in Table III.

Readers are required to perform all the three tests on induction motor and note down the reading in the format shown in Table IV. From the noted readings, calculate the circuit parameters using the mathematical steps explained in section 1.3 and give the results in the format shown in Table V.

No-load test is very important test. It gives information about the magnetic losses (also called as core losses or iron losses), friction and windage losses, no-load power factor, magnetizing current and parameters of the magnetizing branch. No-load loss also reveals any mechanical faults and noise etc. From no-load test conducted on the three-phase induction motor, it can easily be decided, whether the motor is healthy or not. For this, three phase currents (i_a, i_b, i_c) measured during no-load test are mathematically transformed into two phase currents as i_d and i_q as

$$i_d = i_a \qquad\qquad (1.14)$$

$$i_q = \sqrt{2/3}\,(i_b + i_c) \qquad\qquad (1.15)$$

Different set of currents are obtained during no-load test performed with variable voltage to determine and plot the no-load loss at different voltages. Representation of i_d and i_q is a circular pattern centered at the origin of the coordinates. It reveals any abnormal condition due to any fault in the motor by observing deviations in the graph from reference graph of the healthy motor. It can be seen that the plot of i_d and i_q in case of healthy motor is almost in circular shape, and in case of faulty motor, it is more close to an ellipse. Also, the absolute value of current is always higher in case of faulty motor. To perform analysis on the practical data obtained from the measurements, MATLAB software can be used.

While performing the no-load test, the motor runs nearly at synchronous speed i.e. slip is very low or approaching to zero. In this case, the effective rotor resistance is very high, and allows very less current to flow in the rotor circuit. Rotor circuit behaves as open circuit, though the rotor winding is practically closed. This is the reason that no-load test is also termed as open circuit (O.C.) test. Opposite to this, in case of transformer, secondary winding is actually open circuited.

Blocked rotor test cannot be performed on wound rotor motor with open rotor winding (in this case, motor will not run that can be termed as blocked rotor), because with open circuited rotor, no current will flow in rotor circuit. This condition does not comply with those required for the test. Blocked rotor test is performed with low voltage (15-20 % of the rated voltage) applied to stator winding. Also, impedance of the exciting branch is very high as compared to impedance of the rotor circuit. Because of this, very low (almost no

13

current) current pass through exciting branch, and most of the current flow through rotor circuit. Hence, it (magnetizing branch) behaves as open circuited. When the rotor is under blocked condition, slip is 1 that leads to very low rotor resistance, and zero rotor reactance. Under this condition, with open exciting branch and very less rotor resistance, rotor circuit behaves as short circuited. That is why, this test is also termed as Short circuit test.

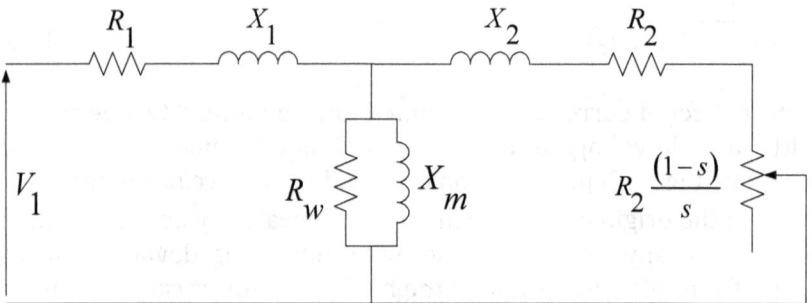

Fig. 1. Per-phase equivalent circuit of three-phase induction motor

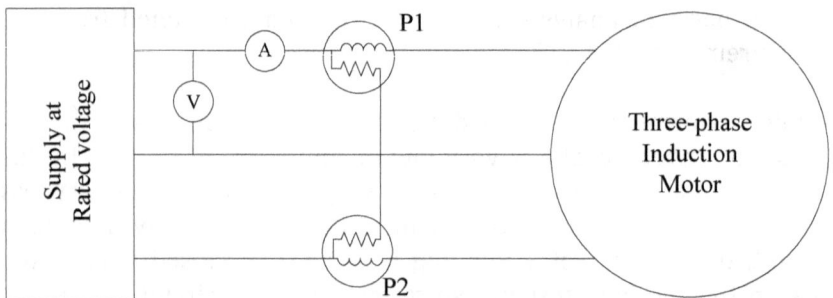

Fig. 2. Circuit connection of no-load test of three-phase induction motor

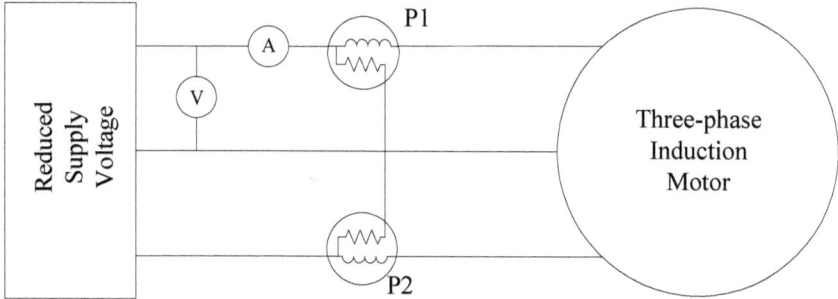

Fig. 3. Circuit connection of blocked rotor test of three-phase induction motor

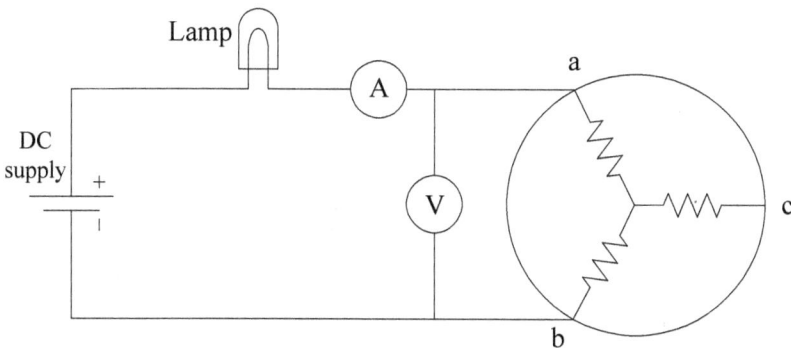

Fig. 4. Circuit connection of DC test of three-phase induction motor

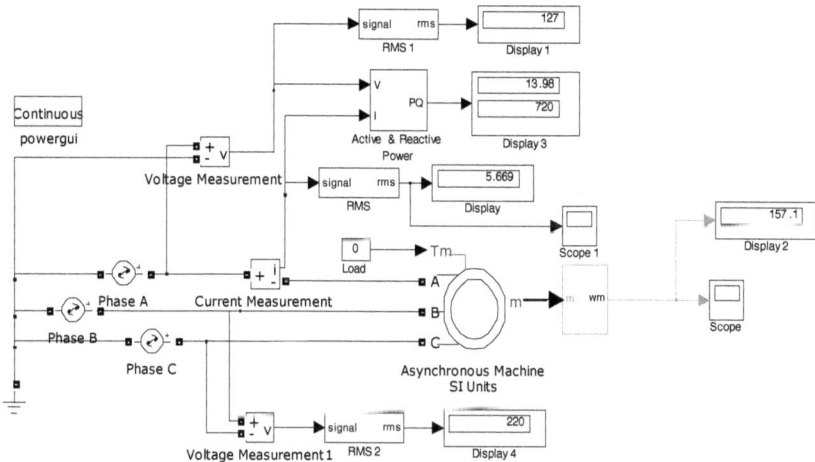

Fig. 5. Simulink model for no-load test of three-phase induction motor

Fig. 6. Setting of motor parameters in machine model

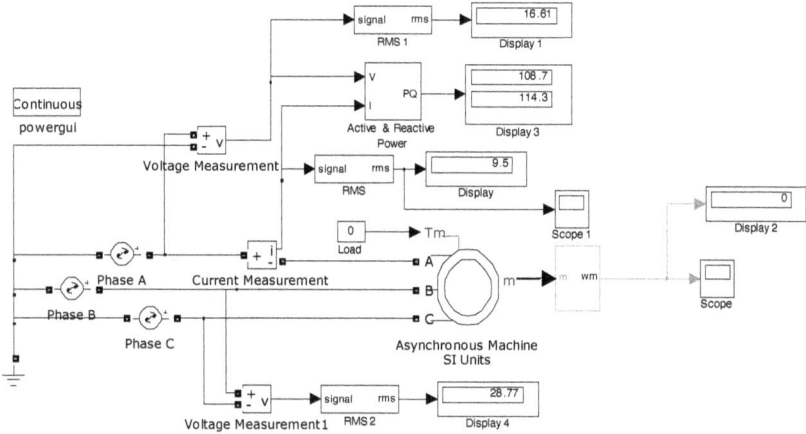

Fig. 7. Simulink model for blocked rotor test of three-phase induction motor

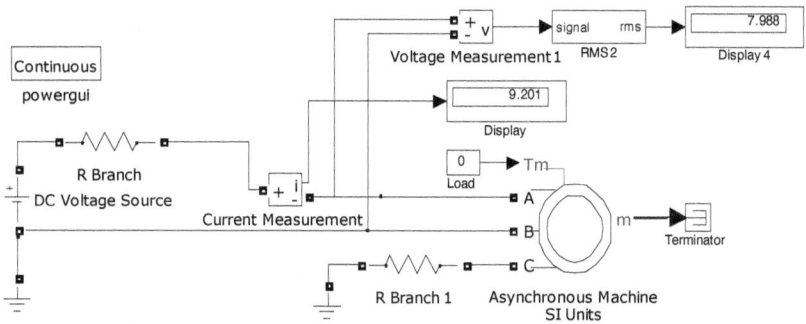

Fig. 8. Simulink model for DC test of three-phase induction motor

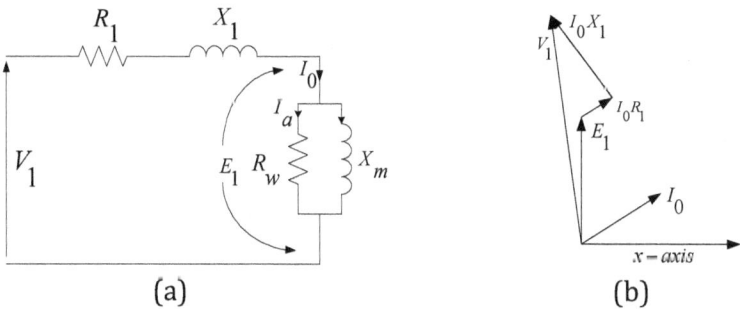

Fig 9. During no-load test (a) per-phase equivalent circuit (b) phasor diagram

17

R_{el} X_{el} $y-axis$

V_{sc} I_{sc}

V_{sc} $X_{el}I_{sc}$

$R_{el}I_{sc}$ I_{sc}

(a) (b)

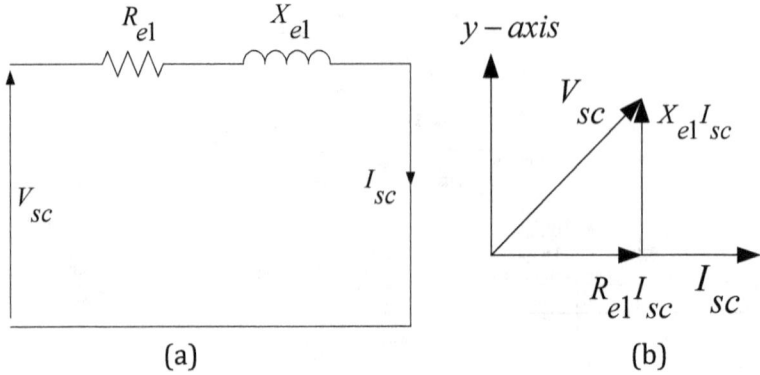

Fig 10. During blocked rotor test (a) per-phase equivalent circuit (b) phasor diagram

Table I: Parameters of three-phase induction motor

$R_1 = 0.435\ \Omega$	$R_2 = 0.816\ \Omega$
$L_1 = L_2 = 2.0\ mH$	$L_m = 234.6\ mH$
$J = 0.089\ Kg.m^2$	

Table II: Observation of no-load and blocked rotor test

Measurement	No-load test	Blocked rotor test	DC test
Voltage	$V_{nl} = 220\ V$	$V_{sc} = 28.77\ V$	$V_{dc} = 7.988\ V$
Current	$I_0 = 5.669\ A$	$I_{sc} = 9.50\ A$	$I_{dc} = 9.201\ A$
Power input	$P_0 = 41.94\ (3 \times 13.98)\ W$	$P_{sc} = 326.10\ (3 \times 108.7)\ W$	

Table III: Test results of three-phase induction motor

DC test	No-load test	Blocked rotor test
$R_1 = 0.438 \, \Omega$	$R_w = 3454.1 \, \Omega$	$R_2 = 0.7662 \, \Omega$
	$X_m = 67.2397 \, \Omega$	$X_1 = 0.6338 \, \Omega$
		$X_2 = 0.6338 \, \Omega$

Table IV: Observation format of three-phase induction motor

Measurement	No-load test	Blocked rotor test	DC test
Voltage	$V_{nl} = ---V$	$V_{sc} = ---V$	$V_{dc} = ---V$
Current	$I_0 = ---A$	$I_{sc} = ---A$	$I_{dc} = ---A$
Power input	$P_0 = ---W$	$P_{sc} = ---W$	

Table V: Result format of three-phase induction motor

DC test	No-load test	Blocked rotor test
$R_1 = ---\Omega$	$R_w = ---\Omega$	$R_2 = ---\Omega$
	$X_m = ---\Omega$	$X_1 = ---\Omega$
		$X_2 = ---\Omega$

EXPERIMENT 2

2.1 Objective: To perform no-load, blocked rotor and DC test on single-phase induction motor.

2.2 Software Required: MATLAB /Simulink software.

2.3 Brief Theory

Operation of single-phase induction motor is based on double revolving theory; according to which, airgap flux is constituted by two equal and oppositely revolving fluxes. Hence, induced electromotive force (emf E) due to resultant airgap flux is constituted by individually revolving fluxes in opposite direction (i.e. E_f: emf due to forward flux direction and E_b: emf due to backward flux direction). This phenomenon is visualized through an equivalent circuit shown in Fig. 1. Motor operation is signified with one stator winding and two imaginary rotors, rotating in opposite direction with each other.

2.3.1 No-load Test

A general circuit connection to perform no-load test of single-phase induction motor is shown in Fig. 2. In this test, motor is allowed to run at rated voltage and frequency under no-load as shown in Fig. 3 (a). In this condition, value of slip is close to zero ($s \approx 0$) and $\frac{R_2}{2s}$ becomes very large compared with $\frac{X_m}{2}$ of forward rotor. In backward rotor, value of $\frac{R_2}{2(2-s)}\left(=\frac{R_2}{4}\right)$ is much smaller than $\frac{X_m}{2}$. Hence, equivalent reactance under no-load is given as

$$X_0 = X_1 + \frac{X_m}{2} + \frac{X_2}{2} \qquad (2.1)$$

From the test, following readings are noted:
Voltmeter reading for input voltage: V_0
Ammeter reading for input current: I_0
Wattmeter reading for input Power: P_0
Hence, no-load power factor is

$$\cos \emptyset_0 = \frac{P_0}{V_0 I_0} \qquad (2.2)$$

No-load impedance

$$Z_0 = V_0/I_0 \tag{2.3}$$

No-load reactance

$$X_0 = Z_0 \sin \varnothing_0 = Z_0 \sqrt{1 - \cos_{\varnothing_0}^2} \tag{2.4}$$

2.3.2 Blocked Rotor Test

During this test with blocked rotor, input voltage of reduced magnitude is fed to flow the current at rated value. Since, rotor is not allowed to rotate so, value of slip $s = 1$. Value of $\frac{X_m}{2}$ is much larger than $\left(\frac{R_2}{2} + j\frac{X_2}{2}\right)$ and is ignored. Equivalent circuit of motor during blocked rotor test is given by Fig. 3 (b). From the test, following readings are noted:

Voltmeter reading for input voltage: V_{sc}
Ammeter reading for input current: I_{sc}
Wattmeter reading for input Power: P_{sc}

Hence, equivalent impedance referred to stator side is

$$Z_{e1} = \frac{V_{sc}}{I_{sc}} \tag{2.5}$$

Equivalent resistance referred to stator side is

$$R_{e1} = R_1 + \frac{R_2}{2} + \frac{R_2}{2} = R_1 + R_2 = \frac{P_{sc}}{I_{sc}^2} \tag{2.6}$$

Therefore,

$$R_2 = R_{e1} - R_1 = \frac{P_{sc}}{I_{sc}^2} - R_1 \tag{2.7}$$

where, R_1 is found by conducting DC test.

Equivalent reactance referred to stator side is

$$X_{e1} = \sqrt{Z_{e1}^2 - R_{e1}^2}$$ (2.8)

Also,

$$X_{e1} = X_1 + \frac{X_{e2}}{2} + \frac{X_{e2}}{2} = X_1 + X_2$$ (2.9)

with consideration,

$$X_1 = X_2 = \frac{1}{2}X_{e1}$$ (2.10)

2.3.3 DC Test

This test is conducted to determine the winding resistance R_1 using voltmeter-ammeter method as shown in Fig. 4. A resistive element (shown as lamp) is connected in series to limit the current in circuit. Care should be taken that current flow must not be more than rated value.

During this test, following observation is noted

Voltmeter reading = V_{dc}
Ammeter reading = I_{dc}
Winding resistance R_1 is given by,

$$R_{dc} = \frac{V_{dc}}{I_{dc}}$$ (2.11)

To consider skin effect, obtained resistance R_{dc} is multiplied by 1.05.

$$R_1 = 1.05R_{dc}$$ (2.12)

2.4 Experimental Procedure using MATLAB/Simulink

In this experiment, a 0.25 HP, 110 V, 50 Hz., single-phase induction motor is selected having the parameters given in Table I.

2.4.1 No-load Test

Step 1: Launch the Simulink in Matlab.
Step 2: In new model file, add the required blocks of 'Single Phase Asynchronous Machine' together with other measuring blocks

22

which are to connected as shown in Fig. 5. Motor parameters are set into machine model through 'Block Parameters' dialog box as shown in Fig. 6.

Step 3: Run the developed test model at rated voltage under no-load condition for some 3 sec.

Step 4: Take the reading of input voltage, current and power, as shown in Table II.

2.4.2 Blocked Rotor Test

Step 1: Simulink model used for no-load test is also used in blocked rotor test as shown in Fig. 7. In the model, rotor is locked by setting the infinite value to moment of inertia.

Step 2: Adjust the input voltage at reduced magnitude for rated current in motor main winding. It may be noted that auxiliary winding is not energized and connected to ground.

Step 3: Take the reading of input voltage, current and power, as shown in Table II.

2.4.3 DC Test

Step 1: Between the any two end terminals of main winding, connect a DC voltage source (50 V in this case), a resistive element (in place of lamp of 10 Ω) along with other measuring units shown in Fig. 8. In the circuit, a resistive element is also added to complete the circuit.

Step 2: Take the reading of input terminal voltage and current, as shown in Table 2.

Step 3: Calculate the value of winding resistance.

2.5 Sample Calculation

Results obtained from tests of three-phase induction motor using mathematical steps are:

A) DC test

$$R_{dc} = \frac{V_{dc}}{I_{dc}} = \frac{8.403}{4.16} = 2.02 \ \Omega$$

$$R_1 = 1.05 * 2.02 = 2.121 \ \Omega$$

23

B) No-load test

Power factor at no-load is given by

$$\cos \Phi_0 = {P_0}/{(V_0 I_0)} = \frac{75.84}{110*3.188} = 0.216$$

$$\sin \Phi_0 = \sqrt{1 - \cos^2_{\Phi_0}} = \sqrt{1 - 0.216^2} = 0.976$$

$$Z_0 = \frac{V_0}{I_0} = \frac{110}{3.188} = 34.5044 \ \Omega$$

$$X_0 = Z_0 \sin \Phi_0 = 34.504 * 0.976 = 33.676 \ \Omega$$

$$X_0 = X_1 + \frac{X_2}{2} + \frac{X_m}{2}$$

where, magnetizing reactance X_m is calculated once leakage reactances X_1 and X_2 are found from blocked rotor test.

C) Blocked rotor test

$$Z_{e1} = \frac{V_{sc}}{I_{sc}} = \frac{35.9}{5.137} = 6.989 \ \Omega$$

$$R_{e1} = \frac{P_{sc}}{I^2_{sc}} = \frac{132.5}{5.137^2} = 5.021 \ \Omega$$

$$R_2 = R_{e1} - R_1 = 5.021 - 2.121 = 2.900 \ \Omega$$

$$X_{e1} = \sqrt{Z^2_{e1} - R^2_{e1}} = \sqrt{6.989^2 - 5.021^2} = 4.862 \ \Omega$$

$$X_1 = X_2 = \frac{X_{e1}}{2} = 2.431 \ \Omega$$

Hence, magnetizing reactance

$$X_m = 2(X_0 - X_1 - \frac{X_2}{2}) = 2(33.676 - 2.431 - \frac{2.431}{2}) = 60.058 \ \Omega$$

2.6 Results and Discussion

From the calculation of the experimental readings, results are given in Table III.

Readers are required to perform all the three tests on induction motor and note down the reading in format shown in Table IV. From the noted readings, calculate the circuit parameters using the mathematical steps explained in section 2.3 and give the results in format shown in Table V.

In revolving field theory, the pulsating sinusoidally distributed m.m.f. in the airgap is divided into two components rotating in opposite directions. Each of these develops torque in opposite directions due to induction motor action. If the motor is made to rotate in a certain direction, the net torque is positive in that direction and it continues to rotate in that direction. With rotor in motion, rotor current induced by backward field is greater than at stand-still and power factor is very low. However, the magnetic effect of rotor current induced by forward field is lower than at stand-still and power factor is high. Hence, with the increase in rotor speed, forward flux wave increases, while backward flux wave decreases. Due to this, with the rotor in motion, the torque of forward field is greater and that of backward is very low. It is worth mentioning here that it is the stator m.m.f. wave that can be split into two counter rotating waves of equal magnitude and not the airgap flux wave. Forward and backward m.m.f. waves are equal in magnitude at any rotor speed, whereas corresponding flux waves are equal at stand-still only. At all rotor speeds, forward flux wave is much greater than the backward flux wave. Because of the presence of backward torque, a single phase induction motor can never attain synchronous speed at no load and even if the losses are negligible.

In case of open circuit (one supply line disconnected) at one of the stator terminals of three-phase induction motor, while it is running, the motor behaves as a single-phase induction motor and continues to rotate with high slip. However, for the same output power, current drawn from the supply will obviously be approximately $\sqrt{3}$ times the primary current and may damage the motor, if allowed to run for a long time.

For blocked-rotor test, the voltmeter used should be of lower range equal to about 15-20% of the rated voltage of the motor, whereas ammeter should have a range about twice the full load current of the motor. For blocked-rotor test, wattmeter should be of high power factor. For no-load /open circuit test, voltmeter should be of range equal to 150% of motor rated voltage, and the ammeter should have a range equal to half of the full-load current of the machine. Wattmeter used for this test should be of low power factor.

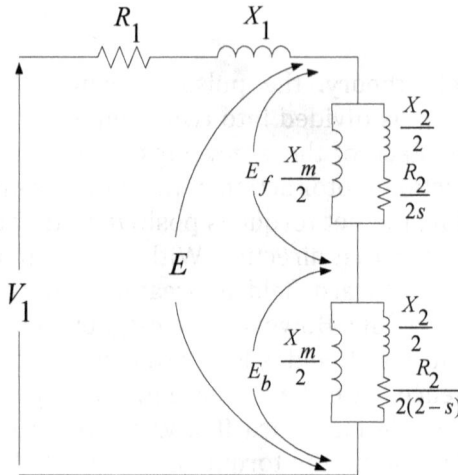

Fig. 1. Equivalent circuit of single phase induction motor

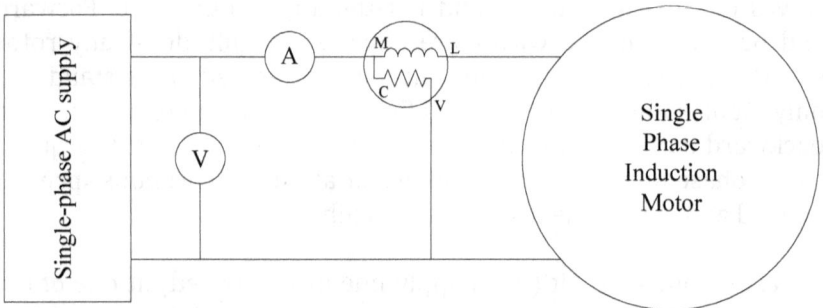

Fig. 2. Circuit connection of single phase induction motor for test

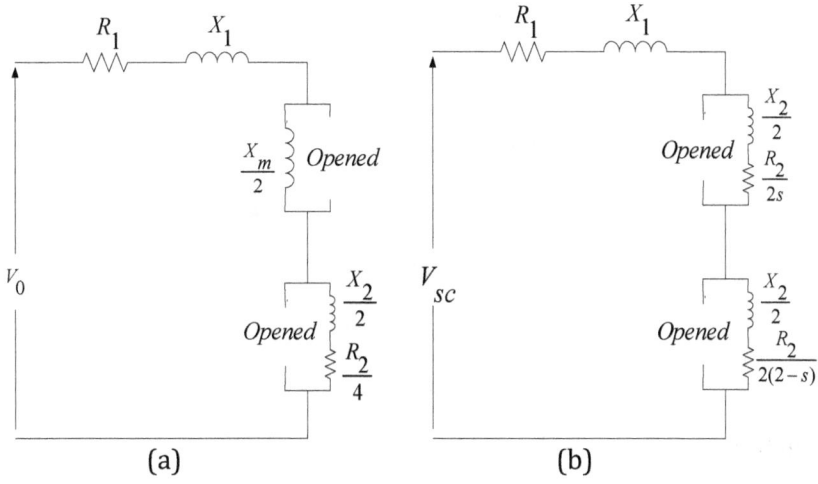

Fig. 3. Equivalent circuit of single phase induction motor under (a) no-load and (b) blocked rotor test

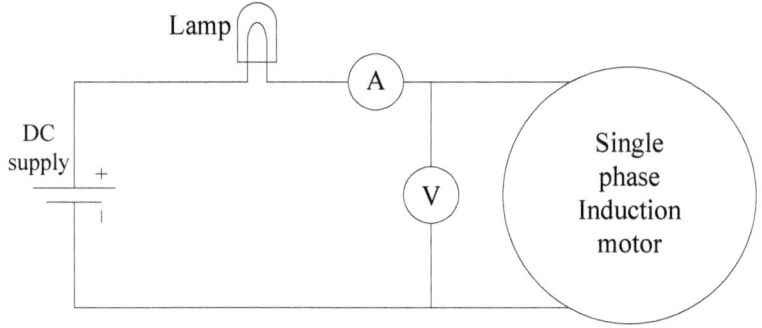

Fig. 4. Circuit diagram of single-phase induction motor for DC test

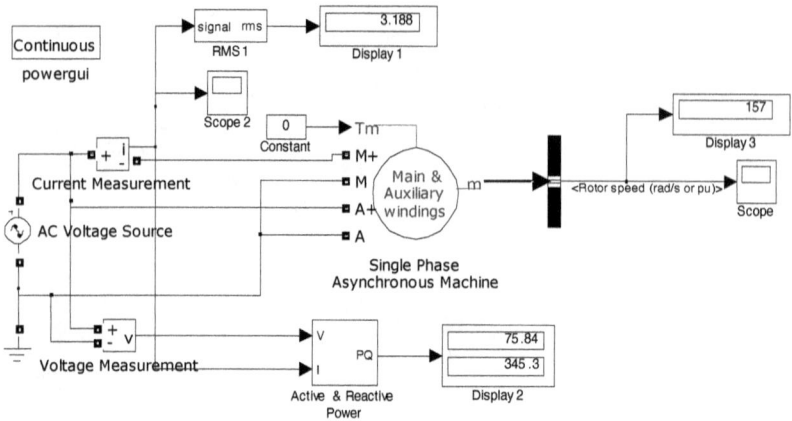

Fig. 5. Simulink model for no-load test of single-phase induction motor

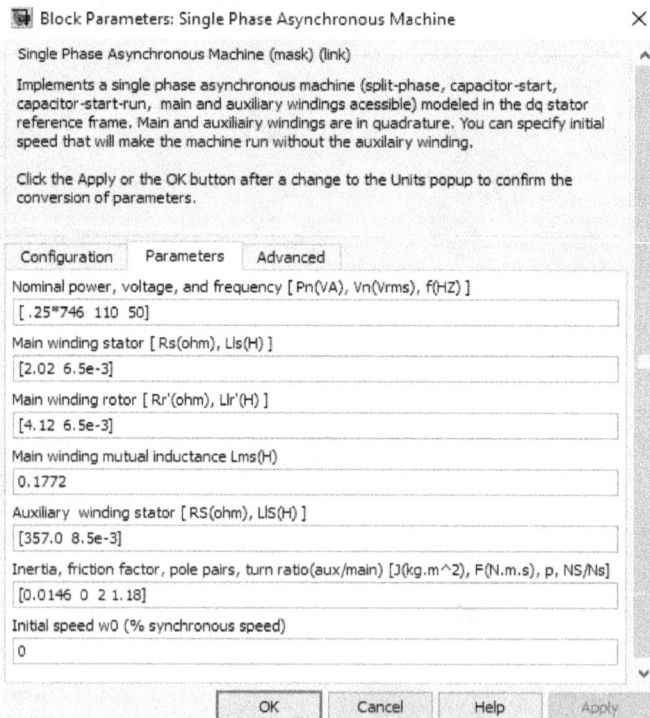

Fig. 6. Setting of motor parameters in machine model

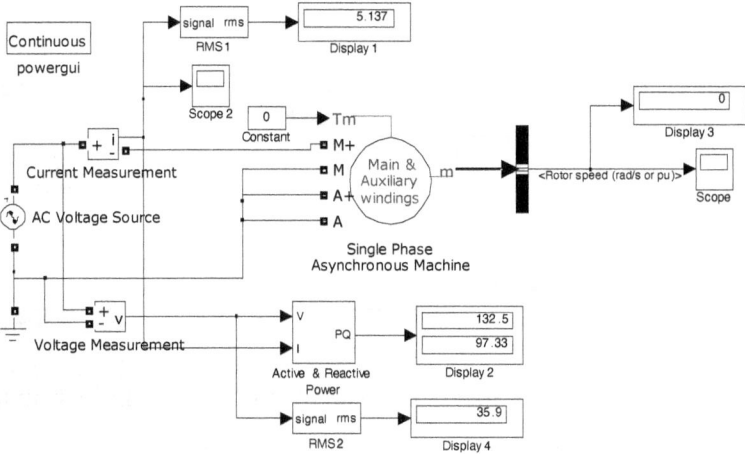

Fig. 7. Simulink model for blocked rotor test of single-phase induction motor

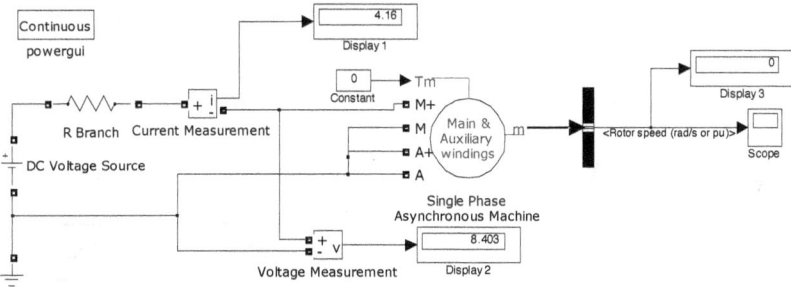

Fig. 8. Simulink model for DC test of single-phase induction motor

Table I: Parameter of single-phase induction motor

$R_1 = 2.02\ \Omega$	$R_2 = 4.12\ \Omega$
$X_1 = 2.36\ \Omega$	$X_1 = 2.36\ \Omega$
$X_m = 55.66\ \Omega$	$J = 0.0146\ Kg.m^2$

Table II: Observation of no-load and blocked rotor test

Measurement	No-load test	Blocked rotor test	DC test
Voltage	$V_0 = 110\ V$	$V_{sc} = 35.9\ V$	$V_{dc} = 8.403\ V$
Current	$I_0 = 3.188\ A$	$I_{sc} = 5.137A$	$I_{dc} = 4.16\ A$
Power input	$P_0 = 75.84\ W$	$P_{sc} = 132.5\ W$	

Table III: Results obtained from tests of single phase induction motor

DC test	No-load test	Blocked rotor test
$R_1 = 2.121\ \Omega$	$X_0 = 33.676\ \Omega$ $X_m = 60.058\ \Omega$	$R_2 = 2.90\ \Omega$ $X_1 = 2.43\ \Omega$ $X_2 = 2.43\ \Omega$

Table IV: Format for observation of readings

Measurement	No-load test	Blocked rotor test	DC test
Voltage	$V_0 = - - - V$	$V_{sc} = - - - V$	$V_{dc} = - - - V$
Current	$I_0 = - - -A$	$I_{sc} = - - - A$	$I_{dc} = - - - A$
Power input	$P_0 = - - - W$	$P_{sc} = - - - W$	

Table V: Format of test results

DC test	No-load test	Blocked rotor test
$R_1 = - - -\Omega$	$X_0 = - - - \Omega$ $X_m = - - - \Omega$	$R_2 = - - - \Omega$ $X_1 = - - - \Omega$ $X_2 = - - - \Omega$

EXPERIMENT 3

3.1 Object: To conduct load test on three-phase induction motor.

3.2 Software Required: MATLAB /Simulink software.

3.3 Brief Theory

Load test is conducted to evaluate the complete performance of three-phase induction motor by determining indices like torque, slip, efficiency etc. In this test, motor is mechanically coupled to a DC generator. In load condition, a variable electrical load (preferably in form of lamp load) is connected to the output of generator. In this set-up, induction motor is driven as prime mover, supplying mechanical power input to the coupled DC generator. When field circuit of coupled DC machine is excited, voltage is generated across the armature terminals of DC machine. Since, armature terminals are connected to external resistive load, power is consumed in it. This power is termed as output power of DC generator. This output power of DC generator is used to obtain its input power (which is the output power of induction motor) To calculate input power of DC generator from its output power, it is essential to calibrate the coupled DC machine for variable speed and field current for its rotational losses and efficiency.

Input power to the motor is measured by using two wattmeter method ($P_{in} = P_1 + P_2$). Complete circuit connection of the experimental setup is shown in Fig. 1.

During this test, following observations were recorded:

Voltmeter reading = V_{in}

Input phase voltage $V_{inp} = V_{in}/\sqrt{3}$ (3.1)

Ammeter reading = I_{inp}

Power input $P_{in} = P_1 + P_2$ (3.2)

Voltmeter reading on generator side = V_{Dc}

Ammeter reading on generator side = I_{DC}

Therefore, output power of generator = $V_{DC} * I_{DC}$ (3.3)

Hence, output power of motor $P_{out} = \dfrac{V_{DC} * I_{DC}}{\eta_{DC}}$ (3.4)

where, η_{dc} is known efficiency of DC generator calculated for given field current and speed. Efficiency of three-phase induction motor is, then determined as

$$\eta_{IM} = \dfrac{P_{out}}{P_{in}} * 100 \qquad (3.5)$$

3.4 Experimental Procedure using MATLAB /Simulink

In this experiment, a 3 HP, 220 V, three-phase induction motor is coupled with a 5 HP, 500/300 V DC generator having the parameters given in Table I and II respectively.

Step 1: Launch the Simulink in Matlab.
Step 2: In new model file, add the required blocks of 'Asynchronous Machine', 'DC Machine' together with other measuring blocks that are to be connected as shown in Fig. 2. Parameters of induction motor and DC generator are set into machine model through 'Block Parameters' dialog box as shown in Fig. 3 (a) and Fig. 3 (b) respectively.
Step 3: In the developed Simulink model, input supply is fed at rated value and operated at no-load. After time t, load torque is switched ON. Input torque T_{eDC} of DC generator is calculated by measured output power of the generator by,

$$T_{eDC} = \dfrac{V_{DC} I_{DC}}{\eta_{DC}\, \omega_r} \qquad (3.6)$$

where, η_{DC} is the known efficiency of DC generator, V_{DC} is armature voltage, I_{DC} is armature current and ω_r is the speed. Efficiency curve of DC generator is shown in Fig. 4. In this figure, an expression was obtained between the generator output power and its efficiency, using curve fitting toolbox. This expression is used to obtain the efficiency of DC generator at a particular output power. Hence, motor torque is calculated by using equation (3.6). Switching of the load torque is achieved by using a switch shown in Simulink model. Fig. 5 shows a dip in speed response of induction motor on switching ON of the load torque at time $t = 8.1$ sec. corresponding to load resistance of 50 Ω at generator output.

Step 4: From the connected Simulink model, note down the input power to induction motor. Per phase input power is shown in the connected block of 'Active & Reactive Power'. Output power of motor P_{out} may be obtained directly from the developed model, using equation (3.4).

Step 5: Calculate the efficiency of three-phase induction motor as shown in observation Table III. Calculate and draw the efficiency and slip curve with respect to output power of motor as shown in Fig. 6 and Fig. 7 respectively. Actual calculated results are plotted. The appropriate curves are shown and mathematically expressed by using 'curve fitting toolbox'.

3.5 Results and Discussion

From the observation and calculation from the experimental readings, performance of three-phase induction motor was evaluated and graphically shown through the curve plot of efficiency and slip curve with respect to output power.

Readers are required to perform all the three tests on induction motor and note down the reading in following format, shown in Table IV. Using the observation and calculation from the experimental readings, draw the performance curve particularly, efficiency and slip curve with respect to output power, as explained above.

The knowledge of rotational loss that constitutes magnetic (hysteresis and eddy current) and mechanical (friction and windage) loss is of great importance to a designer. Poor quality of iron material leads to high hysteresis loss, whereas insufficient degree of lamination and improper insulation between laminations results in increased eddy current losses. Friction and windage losses are dependent on quality of bearing, lubrication and condition of commutator and brush surfaces. Windage loss is caused by air friction. Magnetic loss depends on speed and level of flux in the machine, whereas rotational loss is dependent on speed only.

During calibration of DC machine (determination of iron and mechanical loss), the field current must be adjusted in only one direction (either increasing or decreasing it), so that the point of operation follows the same magnetic path. The DC machine should be allowed to run for sufficient time before recording the readings, to ensure that the brushes and bearings have assumed their normal running condition.

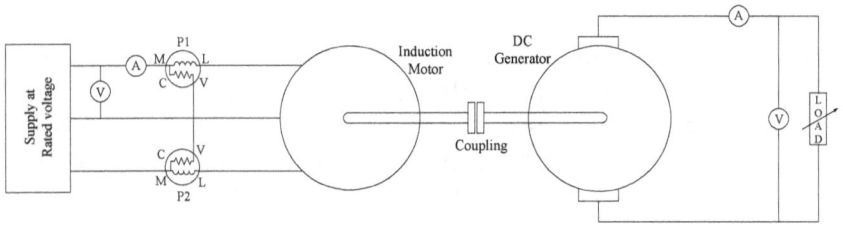

Fig. 1. Circuit connection of load test of three-phase induction motor coupled with DC generator

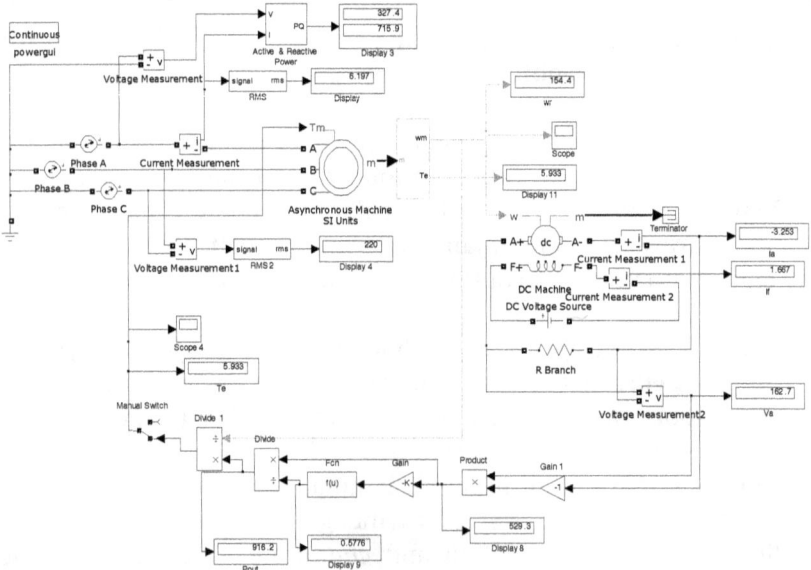

Fig. 2. Simulink model for load test of three-phase induction motor

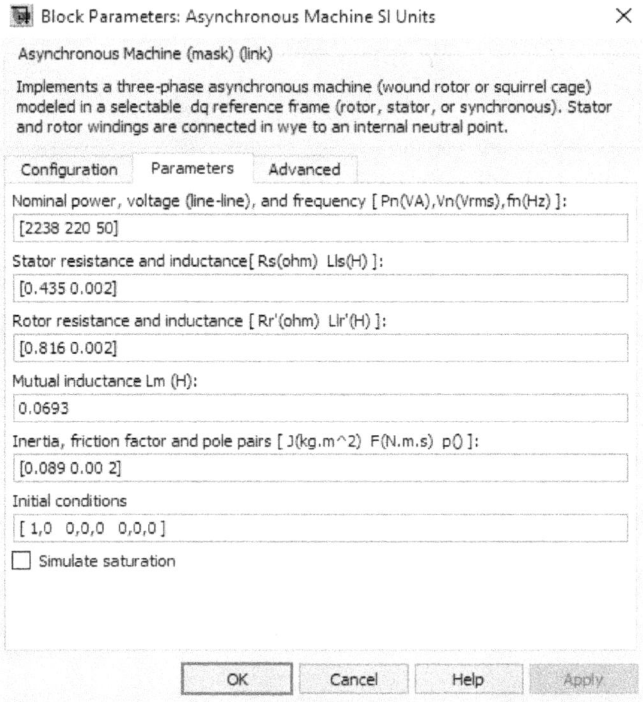

(a)

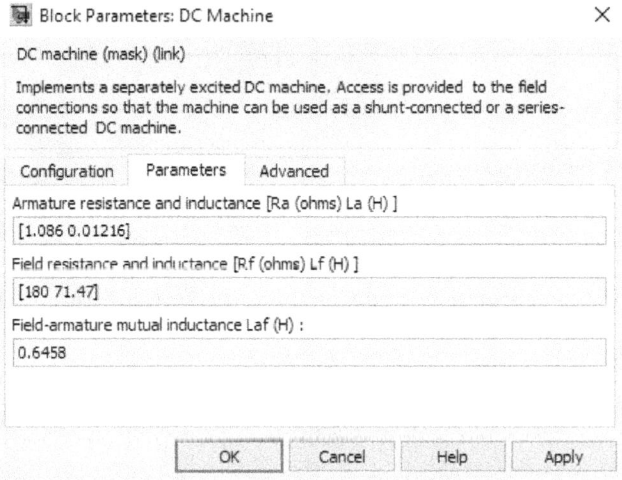

(b)

Fig. 3. Setting of parameters in machine model of (a) induction motor (b) DC generator

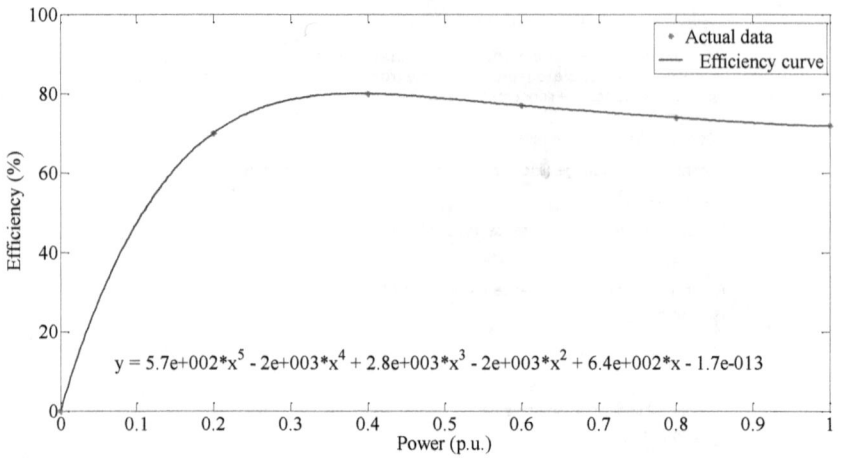

Fig. 4. Efficiency curve of DC generator

$$y = 5.7e+002*x^5 - 2e+003*x^4 + 2.8e+003*x^3 - 2e+003*x^2 + 6.4e+002*x - 1.7e-013$$

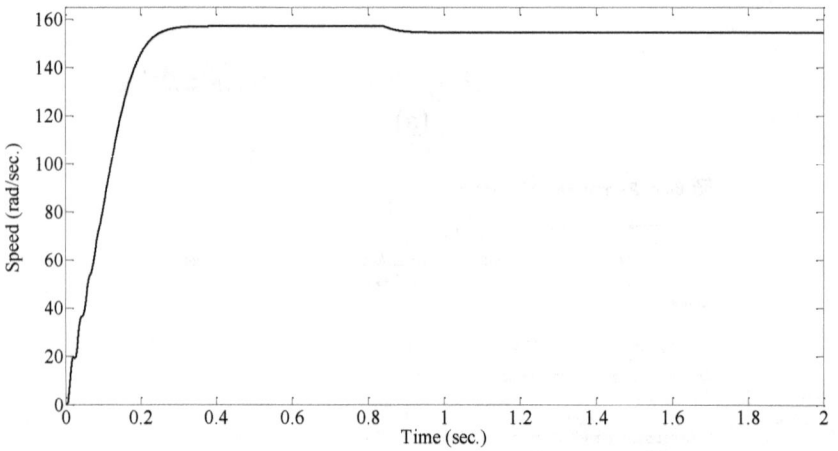

Fig. 5. Speed response of three-phase induction motor

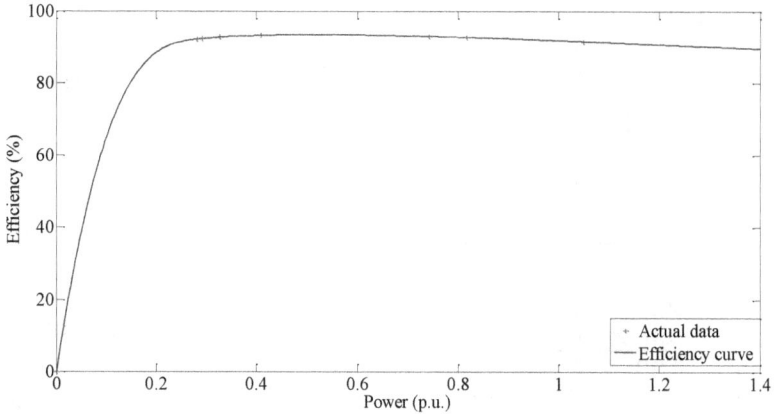

Fig. 6. Efficiency curve of three-phase induction motor

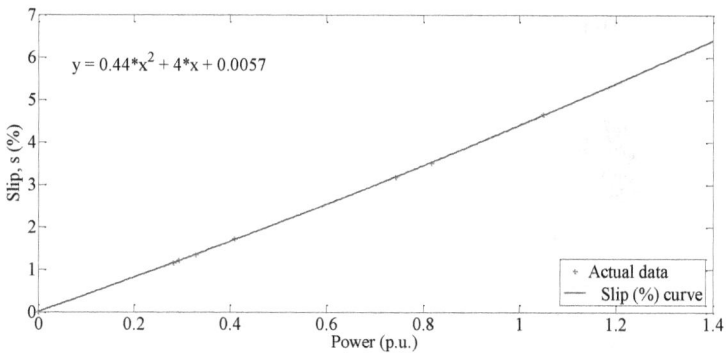

Fig. 7. Slip curve of three-phase induction motor

Table I: Parameter of induction machine

$R_1 = 0.435\ \Omega$	$R_2 = 0.816\ \Omega$
$L_1 = L_2 = 2.0\ mH$	$L_m = 234.6\ mH$
$J = 0.089\ Kg.m^2$	

Table II: Parameter of DC machine

$R_a = 1.086\ \Omega$	$L_a = 0.01216\ H$
$R_f = 180\ \Omega$	$L_f = 71.47\ H$
$L_f = 0.6458\ H$	

37

Table III: Observation of load test of induction motor

S. No.	R load (in ohm)	Speed (rad/sec.)	Input power P_{in} (in Watt)	Output power P_{out} (in Watt)	Efficiency $\eta = \dfrac{P_{in}}{P_{out}} * 100$
1	18	151.6	1972.2	1828.0	92.68
2	20	152.1	1785.0	1660.0	92.99
3	50	154.4	982.2	916.2	93.28
4	100	155.0	791.1	734.0	92.78
5	200	155.2	708.6	654.7	92.39
6	300	155.3	682.8	629.9	92.25

Table IV: Format for observation of load test of induction motor

S. No.	R load (in ohm)	Speed (rad/sec.)	Input power P_{in} (in Watt)	Output power P_{out} (in Watt)	Efficiency $\eta = \dfrac{P_{in}}{P_{out}} * 100$

EXPERIMENT 4

4.1 Object: To study and conduct various starting methods of three-phase induction motor.

4.2 Software Required: MATLAB /Simulink software.

4.3 Brief Theory

At the time of starting, the rotor is at rest (standstill) and the slip is unity ($s = 1$). During this period, the motor behaves as a simple impedance circuit, and since at standstill motor impedance (stator and rotor) is very low, starting current becomes very high of the order of five to seven times the normal full load current. This high current flowing through the stator winding leads to excessive copper loss, and may damage the insulation, Hence, to limit the high starting current, starters are used. In this test, following starting methods are simulated, which are applicable for squirrel cage induction motor.
   a) Direct on-line starter
   b) Star-delta starter
   c) Autotransformer starter

4.3.1 Direct On-Line (DOL) Starter

In this method, lower power rating motor (upto 5 H.P.) is used where it is directly connected to power supply at a particular input voltage as shown in Fig. 1. Because of lower power rating, input current is not very high and motor can withstand during starting time without any starter. In this method, starting torque (T_{st}) is given by

$$T_{st} = T_{fl} \times \left(\frac{I_{sc}}{I_{fl}}\right)^2 \times s_{fl} \tag{4.1}$$

where, starting current is equal to short-circuit current I_{sc}. T_{fl} and s_{fl} indicate full load torque and slip respectively.

4.3.2 Star-delta Starter

In this method, motor is started with stator winding connected in star so that a reduced phase voltage (i.e. $1/\sqrt{3}$ times the line voltage)

39

is applied resulting in reduced starting current. Once, motor pick up the sufficient speed, stator winding is reconfigured to delta by using triple pole double throw switch, as shown in Fig. 2. Hence, starting current is given as

$$\text{Starting line current} = I_{sc}/\sqrt{3} \qquad (4.2)$$

whereas, direct staring current with delta winding $= \sqrt{3}I_{sc}$ (4.3)

$$\text{Hence, } \frac{Starting\ current\ with\ star\ delta\ starter}{Starting\ current\ with\ with\ direct\ switching} = \frac{I_{sc}/\sqrt{3}}{\sqrt{3}I_{sc}} = \frac{1}{3} \quad (4.4)$$

4.3.3 Autotransformer Starter

In this method, three-phase autotransformer is used to supply a re-duced /regulated voltage is applied to stator winding by tapping of transformation ratio x. Hence, starting current is reduced, as shown in Fig. 3.

$$\text{Hence, } \frac{Starting\ current\ with\ autotransformer\ starter}{Starting\ current\ with\ with\ direct\ switching} = x^2 \qquad (4.5)$$

4.4 Experimental Procedure using MATLAB /Simulink

In this experiment, a 1.1 HP, 415 V, three-phase induction motor is selected having following parameters shown in Table I.

4.4.1 Direct On-Line (DOL) Starter

Step 1: Launch the Simulink in Matlab.
Step 2: In new model file, add the required blocks of 'Asynchronous Machine' together with other measuring blocks as shown in Fig. 4. Motor parameters are set into machine model through 'Block Pa-rameters' dialog box as shown in Fig. 5.
Step 3: Run the developed model at no-load with rated input volt-age and plot the response of rotor and input voltage-current, as shown in Fig. 6 and Fig. 7 respectively.

4.4.2 Star-delta Starter

Step 1: Launch the Simulink in Matlab.

Step 2: In new model file, add the required blocks of 'Asynchronous Machine' together with other measuring blocks as shown in Fig. 8 (a).

Step 3: In star-delta starter, reduced input voltage is applied for 1.5 sec. after which, voltage is maintained at rated value. It was realized by using a switch for each input phase. In this switch, second terminal is control input connected to a clock. During starting period, reduced voltage was allowed through third terminal of switch, but after sometime (time $t \geq 1.5$ sec.), rated voltage was allowed through first terminal of switch. Simulated star-delta starter is shown in Fig. 8 (b).

Step 4: Run the developed model and plot the responses of rotor and input voltage-current as shown in Fig. 9 and Fig. 10 respectively.

4.4.3 Autotransformer Starter

Step 1: Launch the Simulink in Matlab.

Step 2: In new model file, add the required blocks of 'Asynchronous Machine' together with other measuring blocks which are to connected as shown in Fig. 11(a). During simulation, two equidistance tappings are assumed between both ends of transformer winding. Hence, reduced voltage was obtained by one-third of rated value. It was simulated by writing a simple program, saved as a function file (file was saved as 'autotf.m'), shown in Fig. 11 (b). Saved function file was called in Simulink model by using block 'Matlab Function'. Simulated model of autotransformer starter is shown in Fig. 11 (c).

Step 3: Run the developed model and plot the responses of rotor and input voltage-current, as shown in Fig. 12 and Fig. 13 respectively.

4.5 Results and Discussion

Three-phase induction motor using different starters was simulated at no-load and motor responses are plotted in above sections. Magnitude of rotor speed, developed torque and per phase stator current (rms value) during starting time (i.e. at $t = 0.2$ and 0.4 sec.) is given in Table II. It may be concluded that autotransformer used as a starter during starting of three-phase induction motor is more efficient in comparison with other considered starter i.e. DOL and Star-delta starter.

Readers are advised to simulate the above discussed starter for three-phase induction motor and plot the various motor responses. Reading during starting time should be also noted in format given in Table III.

Direct-on-line starting is very simple and inexpensive (no separate starter is required) method of starting a cage induction motor. However, in this method of starting, the rate of temperature rise is very high and the motor may get damaged, if the startup period is increased due to excessive load inertia, insufficient rotor resistance and excessive voltage drop in supply mains caused by other connected load.

Autotransformer starting comes under the category of reduced voltage starting of an induction motor. The advantage of this method lies in the fact that the voltage is reduced by transformation and not by dropping the excess voltage in resistor or reactor, and hence the input current and power factor from supply is also reduced compared to stator resistor or reactor starting.

Star-delta starting is applicable for motors, which are designed to run normally with delta connected winding. With star-delta starter, a motor behaves as it is started by an autotransformer starter with voltage transformation ratio $(1/\sqrt{3})$ i.e. 58%. For line voltage higher than 3.3 kV, star-delta starter cannot be used, because in such applications, the motors designed to operate with delta connected stator winding becomes very expensive.

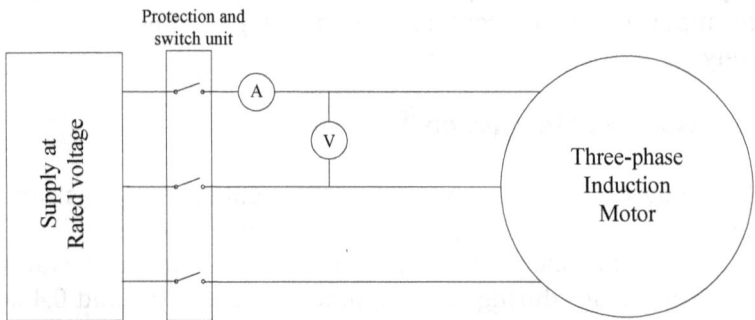

Fig. 1. Direct on-line starting of three-phase induction motor

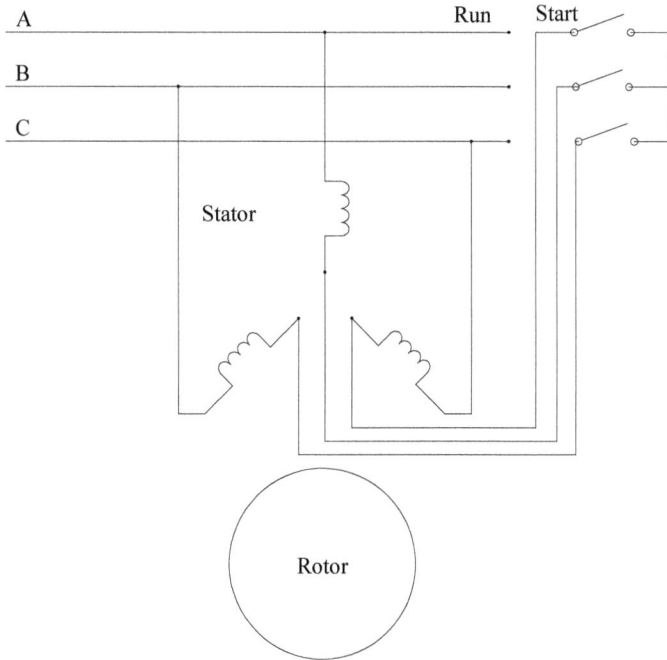

Fig. 2. Starting of three-phase induction motor using Star-delta starter

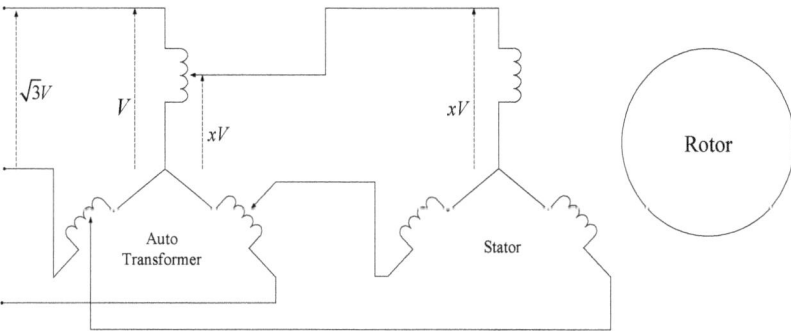

Fig. 3. Auto-transformer starting of three-phase induction motor

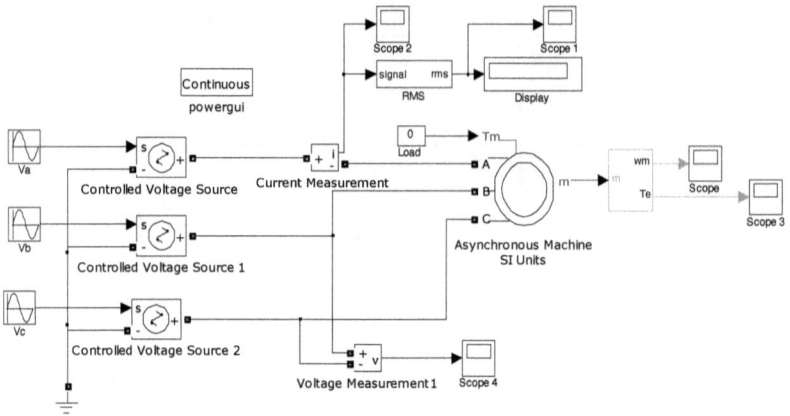

Fig. 4. Developed Simulink model of three-phase induction motor using Direct on-line (DOL) starter

Fig. 5. Setting of motor parameters in machine model

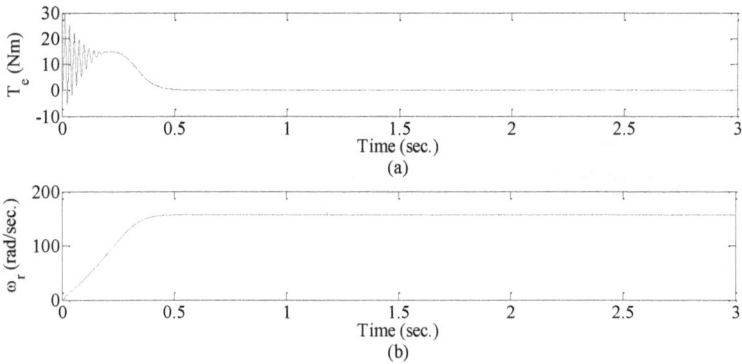

Fig. 6. Motor response with DOL starter (a) developed torque (b) rotor speed

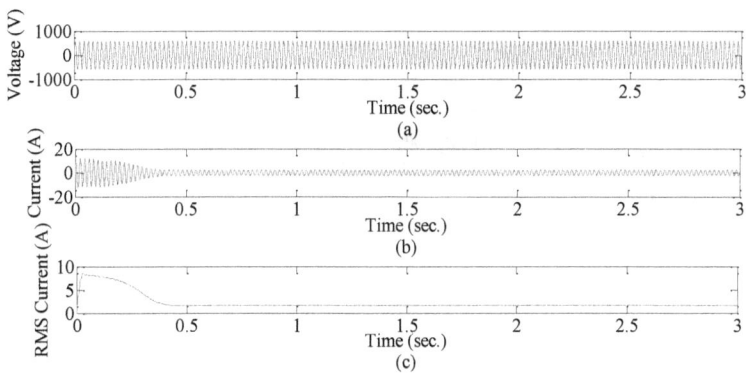

Fig. 7. Voltage-current motor response with DOL starter (a) input phase voltage (b) input phase current (c) rms phase current

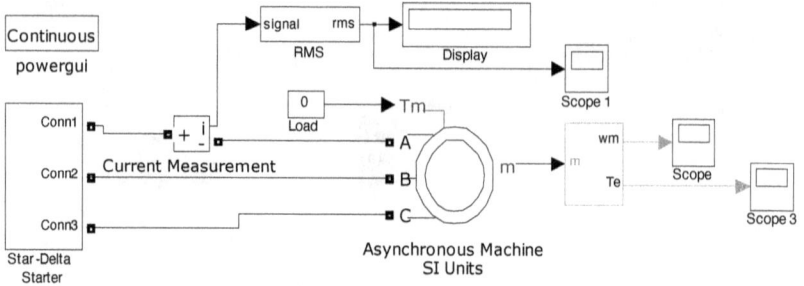

(a)

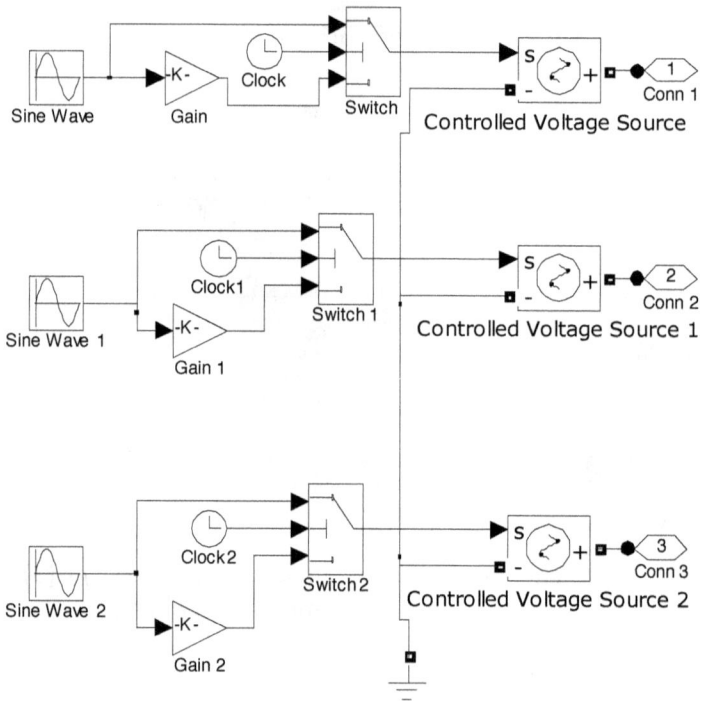

(b)

Fig. 8. Simulink model of three-phase induction motor using starter
(a) complete model (b) star-delta model

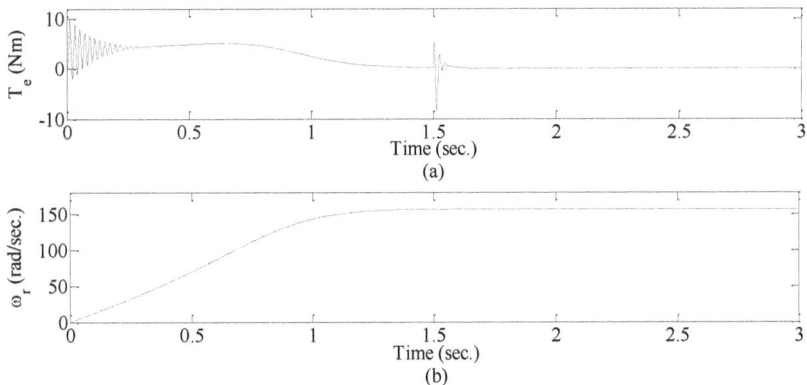

Fig. 9. Motor response with star-delta starter (a) developed torque (b) rotor speed

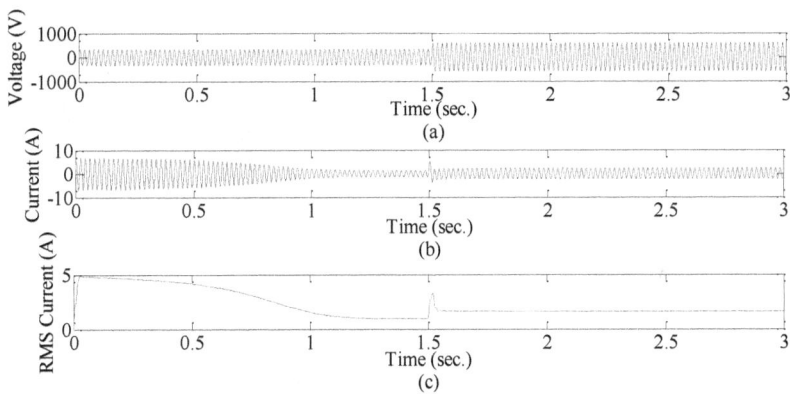

Fig. 10. Voltage-current motor response with star-delta starter (a) input phase voltage (b) input phase current (c) rms phase current

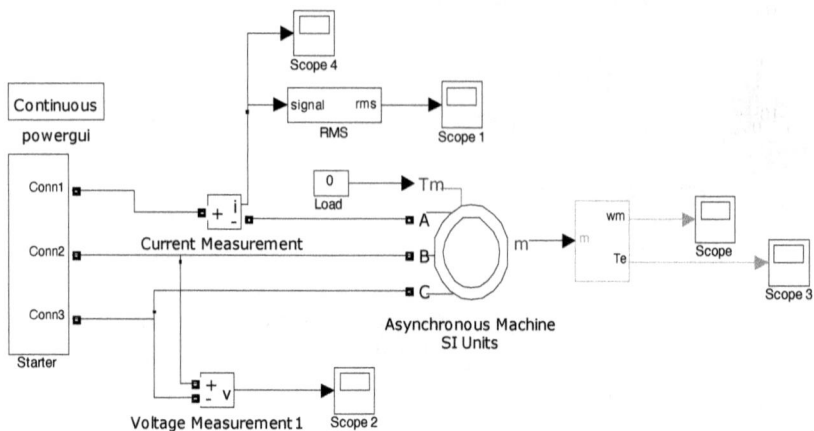

(a)

```
% For autotransformer operation with three tappings
function y = autotf(u)
u=[u(1) u(2)];
if u(1)>=0 && u(1)<0.2;
    y=u(2)/3;
elseif (u(1)>=0.2 && u(1)<0.4)
    y=2*u(2)/3;
else
    y=u(2);
end
end
```

(b)

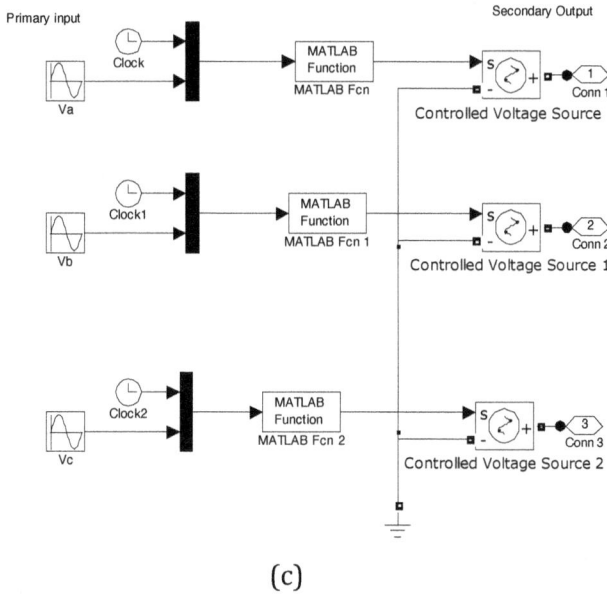

(c)

Fig. 11. Simulink model of three-phase induction motor using auto-transformer starter (a) complete model (b) used program (c) auto-transformer

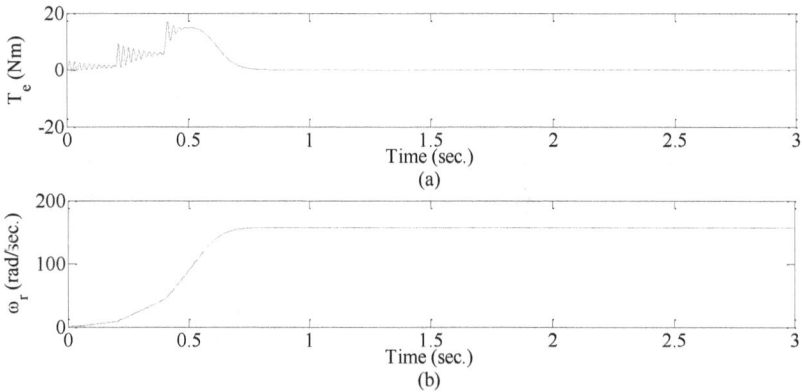

Fig. 12. Motor response with autotransformer starter (a) developed torque (b) rotor speed

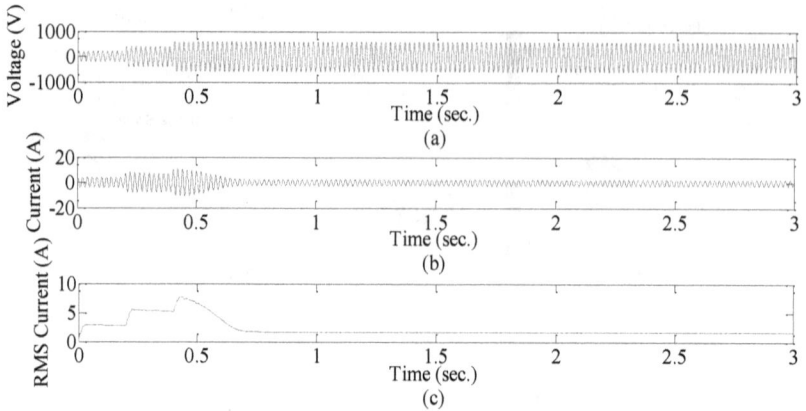

Fig. 13. Voltage-current motor response with autotransformer starter (a) input phase voltage (b) input phase current (c) rms phase current

Table I: Parameters of three-phase induction motor

$R_1 = 6.3\ \Omega$	$R_2 = 10.0\ \Omega$
$L_1 = L_2 = 0.04\ H$	$L_m = 0.42\ H$
	$J = 0.03\ Kg.m^2$

Table II: Motor response with different starters

Starter type	Rotor Speed (rad/sec.)		Torque (Nm.)		Stator current (A)	
	$t = 0.2$	$t = 0.4$	$t = 0.2$	$t = 0.4$	$t = 0.2$	$t = 0.4$
DOL	83.2	155.6	14.7	2.3	6.7	1.8
Star-delta	26.5	55.3	10.0	4.6	4.4	4.6
Autotransform-er	8.2	44.3	1.5	5.8	2.8	5.2

Table III: Format for motor response with different starters

Starter type	Rotor Speed (rad/sec.)		Torque (Nm.)		Stator current (A)	
	$t = 0.2$	$t = 0.4$	$t = 0.2$	$t = 0.4$	$t = 0.2$	$t = 0.4$
DOL						
Star-delta						
Autotransformer						

EXPERIMENT 5

5.1 Objective: To perform polarity test on a single-phase and three-phase transformer.

5.2 Software Required: MATLAB /Simulink software.

5.3 Brief Theory

Polarity test of transformer is conducted to determine the relative direction of induced electromotive force (emf) of two winding transformer. For this purpose, dot convention is adopted for the identification of voltage polarity in transformer. Dot conventions are stated as:

- If the current is entering through dotted terminal of a winding, then the induced voltage on other winding will be positive at the dotted terminal of the second winding.
- If the current is leaving the dotted terminal of one winding, then the polarity of induced voltage in the other second winding will be negative at the dotted terminal.

A two winding transformer is shown in Fig. 1, wherein high voltage (hv) side is indicated with capital letters (A, B, C) and low voltage (lv) side is indicated by small letters (a, b, c). In each side, suffix '1' and '2' show the neutral terminal (if available) and line end respectively.

Polarity test is conducted prior to the parallel connection of two (or more) transformer. In parallel connection, winding terminals with same polarity are connected. Depending on the connection, polarity is classified as

- Additive polarity
- Subtractive polarity

During the polarity test, rated input voltage is fed to one winding (i.e. primary winding), whose an arbitrarily chosen terminal is electrically connected to a terminal of other winding (i.e. secondary winding). A voltmeter is connected between remaining one terminal of both the windings. Polarity is said to be *additive*, if the measured voltmeter reading V_c is more than the input voltage V_a (i.e. $V_c > V_a$). Actual voltmeter reading is given by

$$V_c = V_a + V_b \tag{5.1}$$

where, V_b is the voltage of other winding (i.e. secondary winding) at no-load. Circuit connection showing *additive* polarity of transformer is shown in Fig. 2 (a).

On the other hand, polarity is said to be *subtractive,* if voltmeter reading V_c is less than the input voltage V_a (i.e. $V_c < V_a$). Actual voltmeter reading is given by the potential difference of both the windings i.e.

$$V_c = V_a - V_b \tag{5.2}$$

Circuit connection showing *subtractive* polarity of transformer is shown in Fig. 2 (b).

Polarity test may be conducted for three-phase transformer by adopting the circuit connection for both additive and subtractive for each phase winding on primary and secondary. Detailed experimental procedural steps are given in the following sections:

5.4 Experimental Procedure using MATLAB /Simulink

5.4.1 Polarity Test of a Single Phase Transformer

In order to conduct the polarity test, 1.1 kVA, 230/115 V, single phase transformer (for example) is used and following procedural steps are followed:

Step 1: Launch Simulink in Matlab.
Step 2: Create a new model file, in which all the required blocks (single phase transformer, AC voltage source, voltage measurement etc.) are taken from "SimPowerSystem" library.
Step 3: All the blocks in model file are connected in view of Fig. 2. Developed model of complete system is shown in Fig. 3.
Step 4: Select the suitable solver type and run the developed model. In present simulation, "Variable-step" with 0.0001 sec. as maximum step size, ode 15s (stiff/NDF) solver type was selected.

Step 5: In this developed system, rated single phase voltage (230 V, 50 Hz) is fed to primary winding of transformer at no-load. Terminal voltages (primary voltage V_a and secondary voltage V_b) and voltmeter readings V_c are recorded as shown in observation Table I. In the test, polarity is marked as additive if $V_c > V_a$, and marked as subtractive, if $V_c < V_a$.

5.4.2 Polarity Test of a Three Phase Transformer

In order to conduct the polarity test, 1.1 kVA, 230/115 V, three phase transformer is used and following procedural steps are followed:

Step 1: Launch Simulink in Matlab.
Step 2: Create a new model file, in which all the required blocks (12 terminals, three phase transformer, AC voltage source, voltage measurement etc.) are taken from "SimPowerSystem" library.
Step 3: All the blocks in model file are connected to develop complete system as shown in Fig. 4.
Step 4: Select the suitable solver type and run the developed model. In present simulation, "Variable-step" with 0.0001 sec. as maximum step size, ode 15s (stiff/NDF) solver type was selected.
Step 5: In this developed system, 12 terminal three phase transformer is employed, where end terminals of all three primary and secondary windings are available. Polarity test is conducted on per phase basis. Therefore, single phase voltage (230 V, 50 Hz.) is fed to one primary winding of the transformer at no-load. For one of the transformer phase (say, phase A) terminal voltages (primary voltage V_{A1} and secondary voltage V_{A2}) and voltmeter readings V_c are recorded as depicted in observation Table II. In the test, polarity is marked as *additive*, if $V_c > V_{A1}$, and marked as *subtractive*, if $V_c < V_{A1}$. Similarly, observations are noted for remaining phase B and C, and polarity is decided based on the voltmeter reading.

It is advisable to simulate the developed model to conduct polarity test for both single and three phase transformer of different other ratings, and record their readings in the format shown in Table III and Table IV respectively.

5.4 Results and Discussions

Polarity test was conducted for both single phase and three-phase (on per phase basis) transformer using MATLAB/Simulink. Type of polarity (additive and subtractive) was determined, based on the winding connection.

Parallel operation of transformer with wrong polarity will lead to short-circuit on the transformers. For marking the polarity, the two terminals of the primary winding are arbitrarily marked as terminal 1 and 2. Any one terminal of the primary winding of a single phase transformer (or primary winding of one phase of a three phase transformer) is connected to one terminal of corresponding secondary phase winding of the transformer. A suitable voltage is applied across the primary winding and the reading of the voltmeter connected across the free terminals of primary and secondary is noted down. The connections are then changed /reversed, and the voltmeter reading is recorded with the same voltage applied across the primary terminals. If out of the two cases, the voltmeter gives lower reading with first connection, terminal 1 of the secondary can be marked as 1 and terminal 2 as 2. If the connection 2 gives lower reading, terminal 2 of the secondary is marked as 1 and terminal 1 as 2. It is worth mentioning here that the range of voltmeter should not be less than the sum of the primary and secondary voltages.

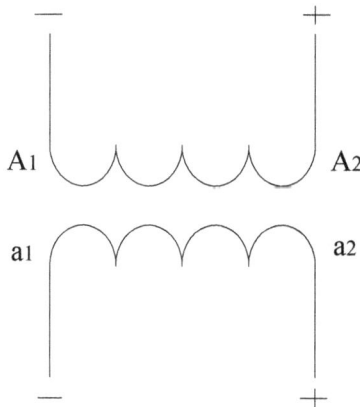

Fig. 1. Labelling of transformer terminals

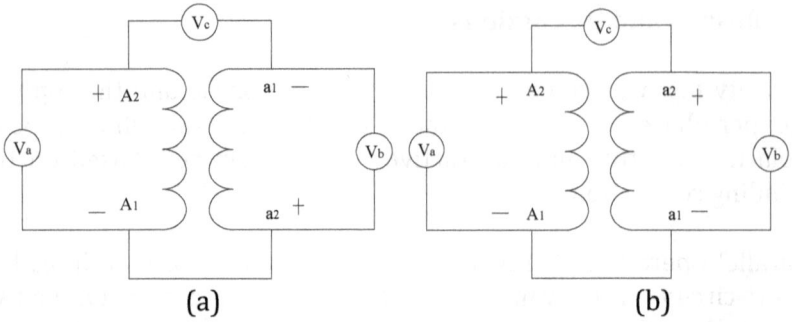

Fig. 2. Polarity test (a) additive (b) subtractive

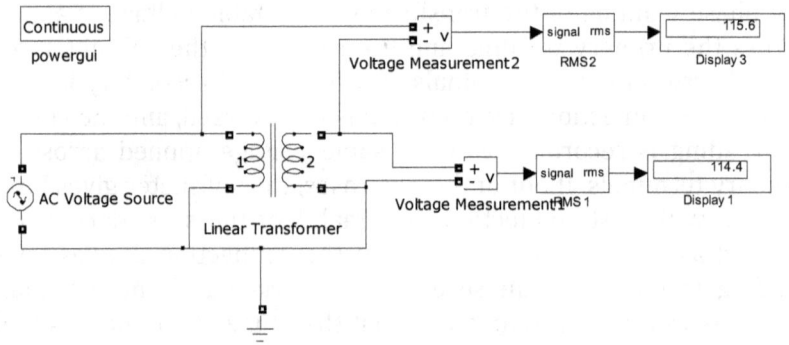

Fig. 3. Model for polarity test of single phase transformer

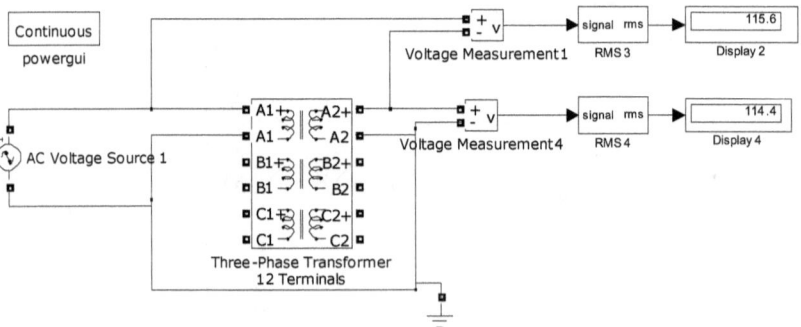

Fig. 4. Model for polarity test of three phase transformer

Table I: Observations for polarity test of single phase transformer

S. No.	V_a	V_b	V_c	Polarity
1.	230.0	114.4	344.4	Additive
2.	230.0	114.4	115.6	Subtractive

Table II: Observations for polarity test of three phase transformer

S. No.	Phase A			
	V_{A1}	V_{A2}	V_c	Polarity
1	230.0	114.4	344.4	Additive
2	230.0	114.4	115.6	Subtractive
	Phase B			
	V_{B1}	V_{B2}	V_c	Polarity
3	230.0	114.4	344.4	Additive
4	230.0	114.4	115.6	Subtractive
	Phase C			
	V_{C1}	V_{C2}	V_c	Polarity
5	230.0	114.4	344.4	Additive
6	230.0	114.4	115.6	Subtractive

Table III: Observations format for polarity test of single phase transformer

S. No.	V_a	V_b	V_c	Polarity

Table IV: Observations format for polarity test of three phase transformer

S. No.	Phase A			
	V_{A1}	V_{A2}	V_c	Polarity
	Phase B			
	V_{B1}	V_{B2}	V_c	Polarity
	Phase C			
	V_{C1}	V_{C2}	V_c	Polarity

EXPERIMENT 6

6.1 Objective: To perform Open-circuit and Short-circuit test on single-phase transformer.

6.2 Software Required: MATLAB /Simulink software.

6.3 Brief Theory

Open-circuit and Short-circuit tests are conducted to determine the circuit parameters, regulation and efficiency of transformer without actual load. In these tests, power consumption is very small, when compared with actual load condition.

6.3.1 Open-circuit Test

Connection diagram for open-circuit test of single phase transformer is shown in Fig. 1. In this test, input voltage is fed to the primary winding (preferably on low-voltage side) at rated value. Secondary side of the transformer is opened circuited and reading is noted by connected voltmeter, ammeter and wattmeter. Ammeter reading gives no-load current I_0. Wattmeter reading gives the iron/core-loss, wherein copper loss (I^2R) is neglected due to small magnitude. During OC test, following readings are noted:

Ammeter reading = no-load current I_0
Voltmeter reading = input rated primary voltage V_i
Wattmeter reading = core loss (P_{oc})

Hence,

$$P_{oc} = V_i I_0 \cos \emptyset_0 \tag{6.1}$$

Power-factor at no-load

$$\cos \emptyset_0 = \frac{P_{oc}}{V_i I_0} \tag{6.2}$$

Active and reactive component of no-load current flowing through resistance R_0 and reactance X_0 are I_a and I_r respectively.

$$I_a = I_0 \cos \emptyset_0 \qquad (6.3)$$
$$I_r = I_0 \sin \emptyset_0 \qquad (6.4)$$

Parameters of magnetizing branch

$$R_0 = \frac{V_i}{I_a} \qquad (6.5)$$

$$X_0 = \frac{V_i}{I_r} \qquad (6.6)$$

6.3.2 Short-circuit Test

Connection diagram for short-circuit test of single phase transformer is shown in Fig. 2. In this test, one side of transformer (preferably low-voltage side) is short-circuited by thick conductor/wire or by connecting ammeter across the winding. Reduced input voltage (about 5 to 10% of rated value) is fed to other side of the transformer. Applied input voltage is adjusted to such a value that a rated full-load current I_{sc} start flowing in the transformer short circuited winding. Reading of the connected voltmeter, ammeter and wattmeter is noted, as shown in Fig. 2.

Voltmeter reading = Input short-circuit voltage V_{sc}
Ammeter reading = Full load rated current I_{sc}
Wattmeter reading = Full load copper loss P_{sc}

Therefore, equivalent resistance referred to primary side

$$R_{e1} = \frac{P_{sc}}{I_{sc}^2} \qquad (6.7)$$

Equivalent impedance referred to primary side

$$Z_{e1} = \frac{V_{sc}}{I_{sc}} \qquad (6.8)$$

Equivalent reactance referred to primary side

$$X_{e1} = \sqrt{Z_{e1}^2 - R_{e1}^2} \qquad (6.9)$$

Power factor,

$$\cos \emptyset_{sc} = \frac{R_{e1}}{Z_{e1}} \qquad (6.10)$$

Also, parameter may be easily calculated referred to other side (secondary side) of tested transformer

$$Z_{e2} = \frac{Z_{e1}}{T^2} \qquad (6.11)$$

$$R_{e2} = \frac{R_{e1}}{T^2} \qquad (6.12)$$

$$X_{e2} = \frac{X_{e1}}{T^2} \qquad (6.13)$$

where, T is transformation ratio of transformer having number of turns N_1 and N_2 in primary and secondary respectively.

6.4 Experimental Procedure using MATLAB/Simulink

In this experiment, a 1.1 kVA, 230/115 V, single-phase transformer is selected having parameters as shown in Table I.

6.4.1 Open-circuit Test

Step 1: Launch the Simulink in Matlab.
Step 2: In new model file, add the required blocks of 'Linear Transformer' together with other measuring blocks as shown in Fig. 3. Transformer parameters are set into the model through 'Block Parameters' dialog box as shown in Fig. 4.
Step 3: Run the developed model with rated input voltage (i.e. 230 V) on primary side at no load with opened secondary terminals. Note down the readings of connected equipment for input voltage, current and power, as shown in observation Table II.

6.4.2 Short-circuit Test

Step 1: Launch the Simulink in Matlab.
Step 2: In new model file, add the required blocks of 'Linear Transformer' together with other measuring blocks as shown in Fig. 5. Transformer parameters are set into the model through 'Block Parameters' dialog box as shown in Fig. 4.

Step 3: Run the developed model with reduced input voltage (5 to 10 % of rated voltage, approximately) on primary side with shorted secondary terminals. Readings on primary side are noted with connected equipment for input voltage, current and power as depicted in observation Table II.

6.5 Sample Calculation

6.5.1 Open-circuit Test

$P_{oc} = 152.2\ W, I_0 = 1.12\ A, V_i = 230\ V$

$$\cos \emptyset_o = \frac{P_{oc}}{V_i I_0} = \frac{125.2}{230*1.12} = 0.591$$

$$\sin \emptyset_o = \sqrt{1 - \cos \emptyset_0^2} = 0.807$$

$$I_a = I_0 \cos \emptyset_o = 0.662\ A$$

$$I_m = I_0 \sin \emptyset_o = 0.904\ A$$

$$R_0 = \frac{V_{oc}}{I_a} = 347.599\ \Omega$$

$$X_0 = \frac{V_{oc}}{I_m} = 254.54\ \Omega$$

6.5.2 Short-circuit Test

$$R_{e1} = \frac{P_{sc}}{I_{sc}^2} = \frac{63.61}{4.79^2} = 2.762\ \Omega$$

$$Z_{e1} = \frac{V_{sc}}{I_{sc}} = \frac{13.5}{4.79} = 2.813\ \Omega$$

$$X_{e1} = \sqrt{Z_{e1}^2 - R_{e1}^2} = \sqrt{2.813^2 - 2.762^2} = 0.534\ \Omega$$

Hence, parameters referred to primary side are calculated from experiment tests and are shown in equivalent circuit in Fig. 6.

6.5.3 Verification of Test

In this test, transformation ratio $t = 2$.

For the transformer used in simulation,

$$R_{e1} = R_1 + t^2 R_2$$
$$= 1.385 + 2^2 * 0.346$$
$$= 2.79 \, \Omega$$
$$X_{e1} = 2\Pi f (L_1 + t^2 L_2)$$
$$= 2\Pi * 50 * (0.000844 + 4 * (0.00021035))$$
$$= 0.534 \, \Omega$$
$$Z_{e1} = \sqrt{R_{e1}^2 + X_{e1}^2}$$
$$= \sqrt{2.79^2 + 0.534^2}$$
$$= 2.835 \, \Omega$$

Hence, a comparison is shown between parameters determined using experimental tests (OC and SC) and that used for simulation as given in Table III.

6.6 Results and Discussion

Open-circuit and Short-circuit test of a single-phase transformer was conducted experimentally on virtual platform using MATLAB/Simulink. Transformer parameters were calculated from experimental test results, referred to primary side. Closed agreement of parameters calculated experimentally and used in simulation validates the correctness of the results.

Readers are advised to simulate the above discussed tests for single-phase transformer and reading should be noted in following format to calculate various parameters, as given in Table IV.

In open circuit test, primary rated voltage is applied at rated frequency with secondary winding open circuited or a voltmeter connected across it. The power input and current are recorded. The no load power input to the transformer is equal to the sum of core loss and copper loss in the primary winding. The latter component is very small as compared to the iron losses, and the no load power input is generally taken to consist of only the iron losses.

In short circuit test, the secondary of the transformer is short circuited and such a reduced voltage is applied across the primary winding that nearly rated current flows in the winding. The power input, primary current and applied voltage are recorded. Under the

63

condition of this test, the applied voltage is very small as compared to normal rated voltage. Hence, the mutual flux in the core is very small, and so iron losses also are very small. Power input in this test is equal to the sum of the copper losses in the primary and secondary windings, neglecting small core losses.

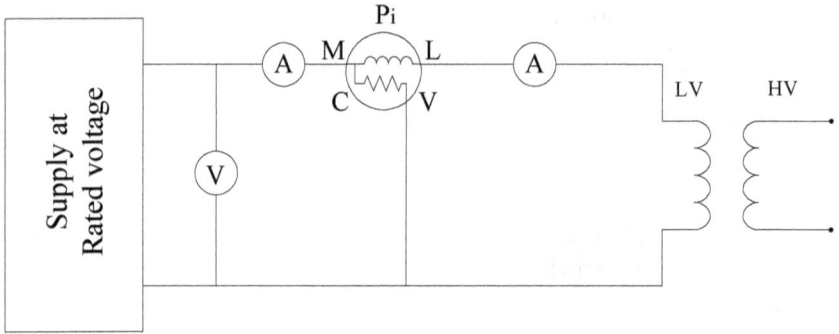

Fig. 1. Connection diagram of open-circuit test of single phase transformer

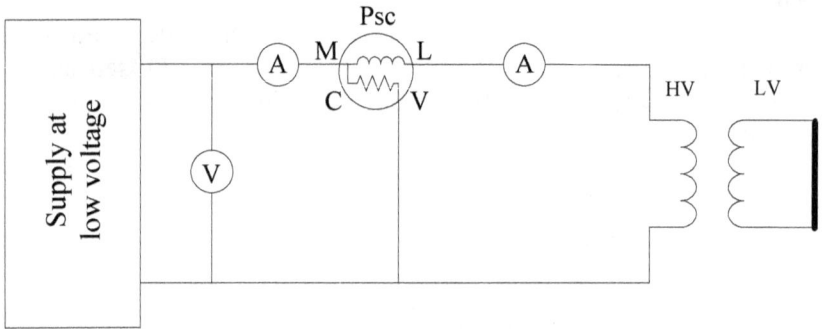

Fig. 2. Connection diagram of short-circuit test of single phase transformer

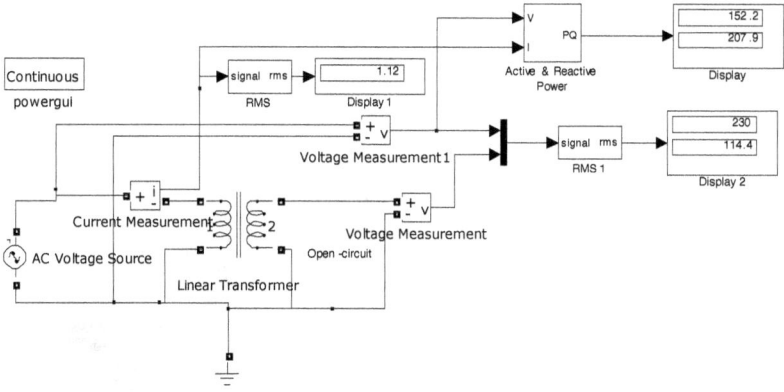

Fig. 3. Developed Simulink model for open-circuit test of single phase transformer

Fig. 4. Setting of transformer parameters for test

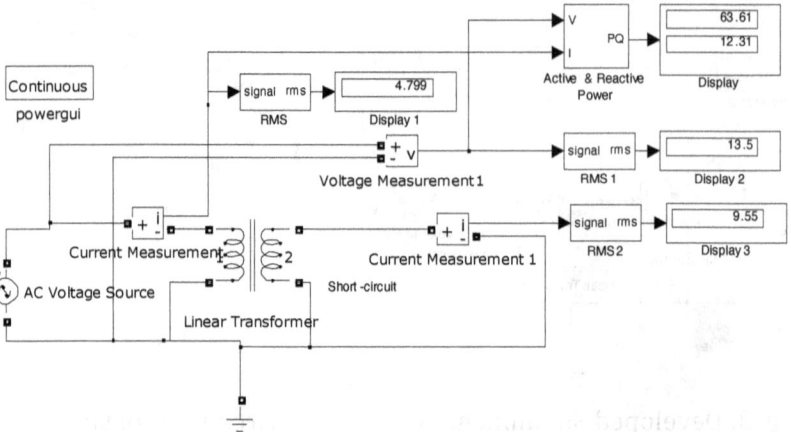

Fig. 5. Developed Simulink model for short-circuit test of single phase transformer

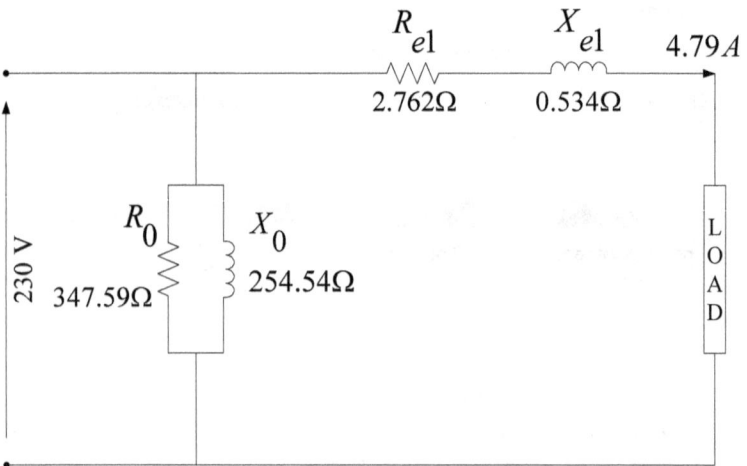

Fig. 6. Approximate equivalent circuit of transformer referred to primary side

Table I: Transformer parameters used during transformer test

Transformer elements	Resistance (Ω)	Inductance (mH)
Magnetizing Branch	348.02	803.09
Primary winding	1.385	0.844
Secondary winding	0.346	0.210

Table II: Observation of transformer test

Type of Test	Voltmeter reading (V)	Ammeter reading (A)	Wattmeter reading (W)
Open-circuit (OC)	230	1.12	152.2
Short-circuit (SC)	13.5	9.55	63.61

Table III: Results comparison

S. No.	Parameters	Experimental results (Ω)	Used transformer parameters (Ω)
1	R_0	347.59	348.02
2	X_0	254.54	252.30
3	R_{e1}	2.76	2.79
4	X_{e1}	0.53	0.53
5	Z_{e1}	2.81	2.84

Table IV: Observation format of transformer tests

Type of Test	Voltmeter reading (V)	Ammeter reading (A)	Wattmeter reading (W)
Open-circuit (OC)			
Short-circuit (SC)			

EXPERIMENT 7

7.1 Objective: To perform Slip Test of Synchronous Machine.

7.2 Software Required: MATLAB /Simulink software.

7.3 Brief Theory

Synchronous machine is considered as an important topic of the courses in electrical machine. In this experiment, salient pole synchronous machine is considered, and synchronous reactances along direct and quadrature axes are determined by conducting slip-test as explained below:

Slip Test of Synchronous Machine

Direct axis synchronous reactance X_d is the reactance of armature coil offered to the flow of direct axis current (i_d). It is defined as the ratio of the fundamental component of reactive armature voltage due to the fundamental direct axis component of armature current to this component of current under steady state condition and at rated frequency. Quadrature axis synchronous reactance X_q is the reactance of armature coil offered to the flow of quadrature axis current (i_q). It is defined as the ratio of the fundamental component of reactive armature voltage due to the fundamental quadrature axis component of armature current to this component of current under steady state condition and at rated frequency.

Synchronous reactance (X_d and X_q) along direct and quadrature axes are determined by conducting slip test at no load. A systematic connection diagram is shown in Figure 1. A reduced three phase voltage (25% of rated value) at rated frequency is used as input to feed armature winding. Field winding is kept open (or connected with a voltmeter). Rotor of the machine is rotated at a slightly less or more than the synchronous speed through a mechanically coupled prime mover (usually a DC shunt machine). During running condition, maximum and minimum value of connected voltmeter and ammeter readings (V_{min}, V_{max} and I_{min}, I_{max}) are recorded. A correct rotational direction of rotor is signified by a small voltage induced in field winding. Rotor is running at speed N_r closed to rotating armature flux i.e. synchronous speed N_s. At some instant, armature flux mmf at slip speed $(s = N_s - N_r)$ will pass through the actual silent field pole. After a quarter of slip cycle, armature flux mmf will be in line with the axis of actual silent field pole (i.e. inter-polar rotor axis). Along d-axis, air-gap is minimum and reluctance offered will be minimum, hence minimum value of magnetizing current I_{min} is indicated by ammeter connected to armature. Along q-axis, air-gap is maximum and reluctance offered will be maximum, hence maximum value of magnetizing current is indicated by amme-ter connected to armature. A typically captured view of the oscillo-scope voltage-current waveform is shown in Figure 2. Ratio of rec-orded magnitude value of voltage and current along d-q axes yields synchronous reactance X_d and X_q:

$$X_d = \frac{V_{max}}{\sqrt{3}I_{min}} \tag{7.1}$$

$$X_q = \frac{V_{min}}{\sqrt{3}I_{max}} \tag{7.2}$$

During slip test, slip should be small as possible to prevent the er-rors in measurement due to induced voltage in damper windings i.e. a lower value of synchronous reactance will be obtained at larger slip. But, at a very low slip, reluctance torque is developed due to saliency tending to bring the rotor to align with the rotating arma-ture mmf wave. Hence to reduce the reluctance torque, slip test is conducted at reduced armature voltage. Furthermore, with constant input voltage, a small variation in terminal voltage is also noted due to voltage drop in connecting wires caused by the fluctuation of cur-rent (as noted in ammeter readings).

7.4 Steps to Perform Slip-test

Following essential steps are adopted to virtually conduct the slip test of synchronous machine using MATLAB/Simulink. For this pur-pose, a three-phase synchronous machine was selected, having the parameters as given in Table I.

Step 1: Launch Simulink in MATLAB.
Step 2: Create a new model file, in which the required blocks are picked-up from different components. All component blocks may be picked from the libraries of Simulink and SimPowerSystem.

Step 3: In view of the connection diagram, component blocks are connected to develop the model to simulate slip-test, as shown in Figure 3. Input terminal voltage with reduced magnitude (100 *V*) is supplied to the test machine.

Step 4: After setting a suitable solver, the developed model is run at a speed (i.e. 150 rad/sec.) slightly less than the synchronous speed. In this experiment, ode 4 solver type (Runge Kutta) was selected with 0.0001 sec. as its step size.

Step 5: Response waveforms of terminal voltage-current are plotted and the readings are noted in observation Table II.

It is worthwhile to mention here that in terminal voltage no variation was noted. This is because of negligible voltage drop in connecting wiring. To ensure higher accuracy in obtained results, it is advisable to observe and record more readings at different rotor speed and/or voltage and take the average value of evaluated reactance/inductance.

From the noted readings, obtained average value of evaluated magnetic reactance/inductance along *d-q* axes are

$X_d = 35.22 \ \Omega$

$L_d = \frac{35.22}{(2\pi f)} = 0.122 \ H$ at *50 Hz.*

$X_q = 15.51 \ \Omega$

$L_q = \frac{15.51}{(2\pi f)} = 0.0494 \ H$ at *50 Hz.*

Hence, the value of magnetizing inductance obtained experimentally was found to be in a close agreement to rated machine parameter.

7.5 Results and Discussion

Slip test of a three-phase synchronous machine was conducted by using MATLAB/Simulink and reactance along direct and quadrature axis was determined. Readers are suggested to consider three-phase synchronous machine with different parameters and note the readings in the format shown in Table III to calculate reactance X_d and X_q along direct and quadrature axis respectively.

The important precaution for this test is to keep the field circuit closed unless and until the slip is small enough, otherwise the induced voltage in the open field winding may reach dangerously high, when slip is substantial. The switch connected across the field winding should be opened only, when it is ensured that the slip is very small. During this test, voltage induced in across the field winding is of slip frequency and has its maxima position along V_{min} of the applied voltage. However, due to residual magnetism of the field system, the field voltage maxima and I_{max} may not coincide with V_{min}. This test is suitable only for determination of X_q. The value of X_d is normally obtained from open circuit (O.C.) and short circuit (S.C.) test.

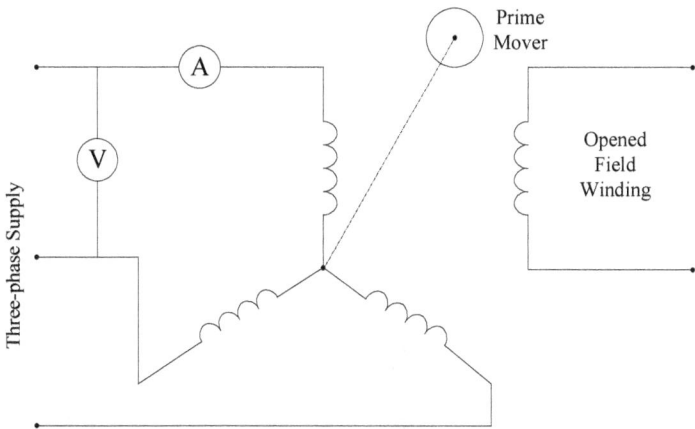

Fig. 1. Connection diagram for slip test

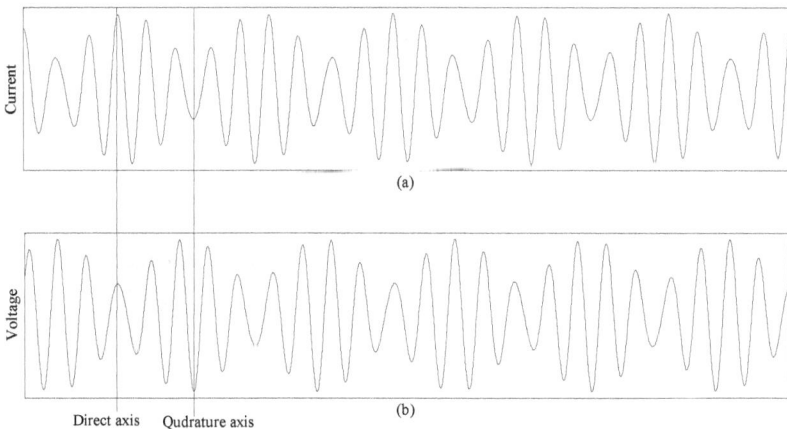

Fig. 2. Typical oscilloscope pattern of waveform (a) phase current (b) phase voltage

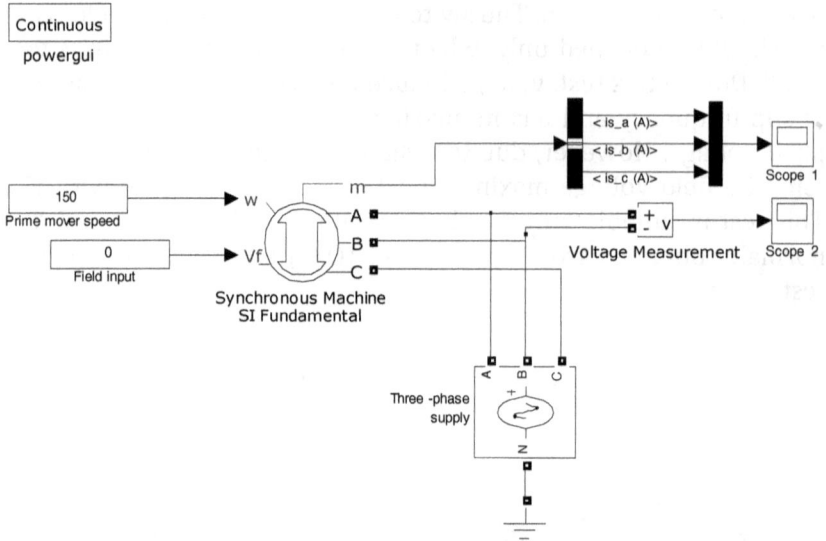

Fig. 3. Developed simulation model for slip-test of synchronous machine

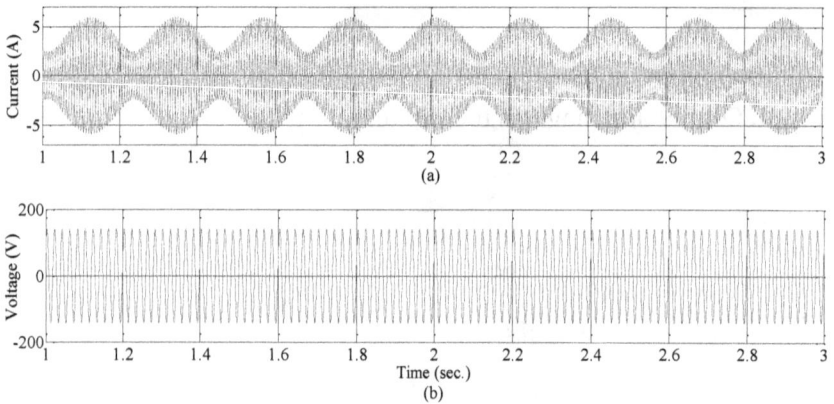

Fig. 4. Voltage-current waveform during slip-test (a) three-phase armature current (b) terminal voltage

Table I: Parameters of 8.1 kVA, 400 V, 1500 rpm, three-phase synchronous machine

S. No.	Parameters	Value
1	Stator resistance R_s	1.62 Ω
2	Stator leakage inductance L_{ls}	4.527 mH
3	Magnetizing inductance L_{mq}	51.75 mH
4	Magnetizing inductance L_{md}	108.6 mH
5	Field resistance R_f	1.208 Ω
6	Field leakage inductance L_{lfd}	11.32 mH
7	Damper resistance R_{kd}	3.142 Ω
8	Damper resistance R_{kq}	4.722 Ω
9	Damper leakage inductance L_{lkd}	7.334 mH
10	Damper leakage inductance L_{lkq}	10.15 mH

Table II: Observation of slip-test

S. No.	Current magnitude (A)		Rotor speed (rad/sec.)	Reactance	
	Minimum value (I_{min})	Maximum value (I_{max})		X_d (Ω)	X_q (Ω)
1	2.41	5.94	150	33.88	13.75
2	2.30	5.31	152	35.50	15.37
3	2.25	4.69	155	36.29	17.41
Average value				35.22	15.51

Table III: Observation format of slip-test

S. No.	Current magnitude (A)		Rotor speed (rad/sec.)	Reactance	
	Minimum value (I_{min})	Maximum value (I_{max})		X_d (Ω)	X_q (Ω)
Average value					

EXPERIMENT 8

8.1 Objective: To perform synchronized operation of synchronous generator.

8.2 Software Required: MATLAB /Simulink software.

8.3 Brief Theory

In power system, a large number of parallel connected alternators are operated simultaneously in synchronism with each other, resulting in the following potential advantages:

i.   A single generator may not be capable to supply load demand of larger magnitude. Such load is supplied by several connected synchronized operated generator,

ii.   During low load demand, running alternator is operated near full load condition (i.e. at higher efficiency) by shutting down some alternators.

iii. During fault condition, remaining operational alternators are capable to supply the load demand, signifying the reliability of system.

iv.   With increasing demand in future, more generators may be added in installed system with no any major modification.

v.   With such a system, energy cost is low to the load demand.

Process of connecting to alternators or synchronous generators in parallel under voltage, without any interruption is called synchronization. The synchronous machine, which is to be connected or synchronized is called incoming machine. Hence, to obtain a synchronized operation of incoming generator with another operating alternator (or infinite bus bar), certain conditions must be fulfilled. These conditions are:

i.   Terminal voltage of incoming alternator and the already operating alternator or bus bar must be equal.

ii.   Frequency of generated voltage of incoming machine and that of the running alternator or bus bar must be nearly equal in magnitude. Practically, frequency of the incoming machine must be slightly more at the time of synchronization.

iii. The two voltages must be in same phase with respect to the external load i.e. phase sequence of two voltages must be the same. If the frequency/speed of incoming machine is less

74

than that of already running alternator, incoming machine starts working as a synchronous motor soon after the synchronizing switch is closed.

A general circuit diagram for synchronization of incoming generator 2 with running generator 1 connected to a load is shown in Fig. 1.

8.4 Steps to Perform Experiment

In order to obtain a synchronized operation of generators, three-phase, 8.1 kVA, 4-poles two identical synchronous machines are used with their parameters mentioned in Table I. For simplicity, rotor speed variation is neglected, and following steps are performed:

Step 1: Launch Simulink in Matlab.
Step 2: Create a new model file, in which all the required blocks (synchronous machine, three-phase breaker, load etc.) are taken from "SimPowerSystem" library.
Step 3: All the blocks in model file are connected in view of Figure 1. Developed model of complete system is shown in Fig. 2.
Step 4: In this developed system, field circuit was excited with DC source at 12 V and generator is connected to an arbitrary prime mover with constant rotating speed at 157.08 rad/sec. Suitable parameters are fed into the blocks through their "Block parameters" dialogue box. It is shown for three-phase synchronous machine and breaker circuit in Fig. 3 (a) and Fig. 3 (b) respectively. It may be noted that at no-load condition, voltage and frequency of generator 1 and generator 2 both will be identical.
Step 5: Select the suitable solver type and run the developed model. In present simulation, "variable step" with 0.001 sec. as maximum step size, ode 15 solver type was selected.
Step 6: Responses of alternators are recorded and displayed.

Prior to operate the three-phase breaker to ON position, all the necessary conditions of synchronization should be satisfied. Initially, running generator 1 is connected to 4000 W active load and current flowing in each phase is 4.6 A (rms value). Breaker is operated at time t = 1 sec. to connect incoming generator 2 for a synchronized operation with running alternator i.e. generator 1. Load sharing may be noted between generator 1 and generator 2 with a simultaneous variation in current i.e. a decrease and increase to 2.74 A (rms value) respectively. Dynamic response of current is shown in Fig. 4. Current waveform of generator 1 during steady-state before and after the operation of breaker is shown in Fig. 5 (a) and Fig. 5 (b) respectively. Whereas, current waveform of generator 2 during steady-state after the operation of breaker is shown in Fig. 5 (c). Voltage and phase angle differences vanish after a synchronized operation and are shown in Fig. 6 (a) and Fig. 6 (b) respectively.

Readers are advised to also simulate the developed system for synchronized operation of synchronous generator with different operating conditions (load, frequency and voltage).

8.5 Results and Discussion

In this experiment, an incoming synchronous generator was operated in synchronism with other alternator by using MATLAB/Simulink. During steady-state, equal power was constituted by incoming synchronous generator and alternator (as both the machines are identical). Readers are advised to repeat this experiment by using the other synchronous machines, which may/may not be identical.

The first condition for synchronization can be fulfilled by a voltmeter, and it is desirable to use the same voltmeter for measuring both the source voltages. All the three conditions, for successful synchronization, can be checked by the use of synchronizing lamps or by synchroscope in association with phase sequence indicator.

The parallel operation of synchronous machines is made possible by the fact that the armature impedances are predominantly reactive. When only two machines of comparable capacity are operating in parallel, a change in the operating conditions of one of them changes both the bus bar voltage and the frequency.

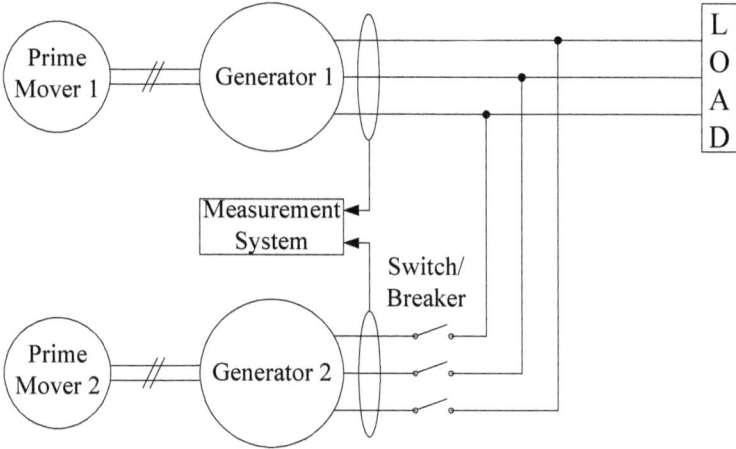

Fig. 1. Synchronization of incoming generator 2

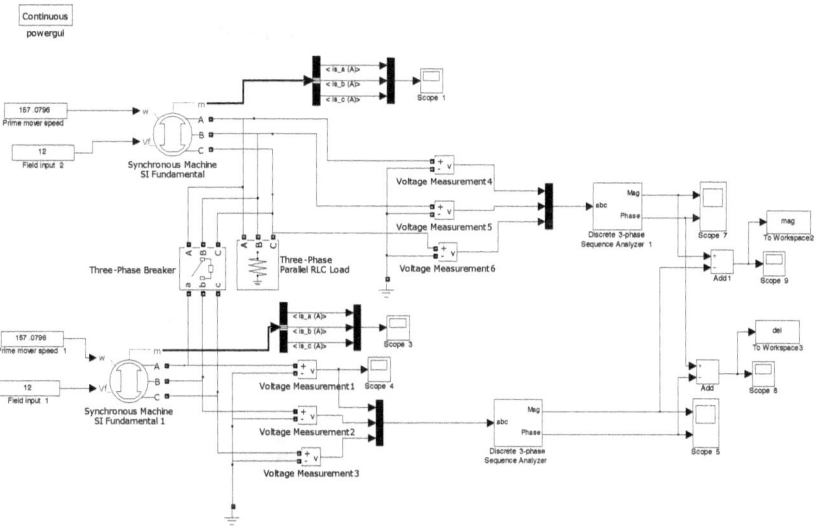

Fig. 2. Simulink model of synchronized generator operation

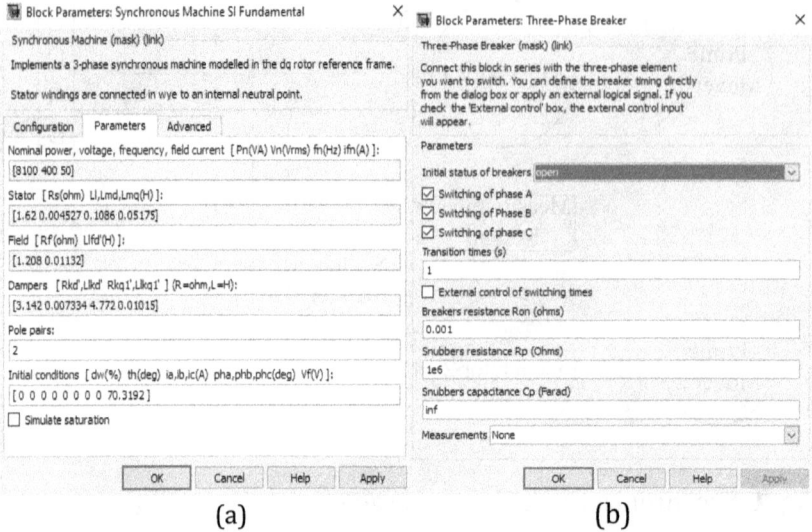

Fig. 3. "Block parameters" dialog box (a) three-phase synchronous machine (b) three-phase breaker

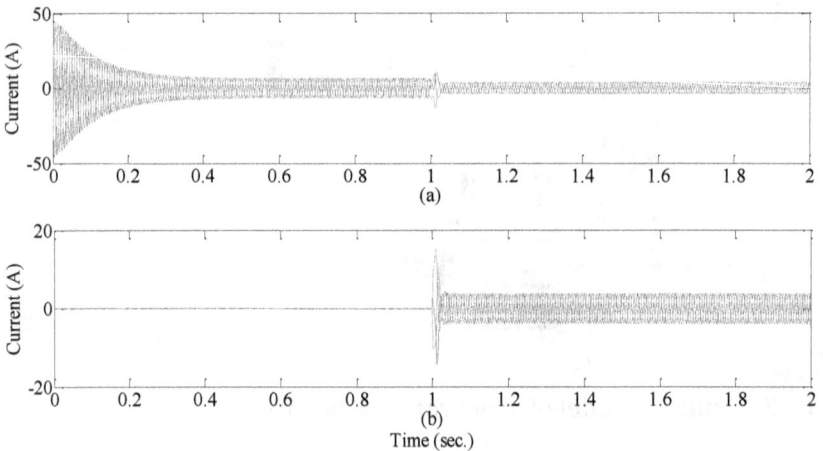

Fig. 4. Dynamic response of current of (a) running generator 1 (b) incoming generator 2

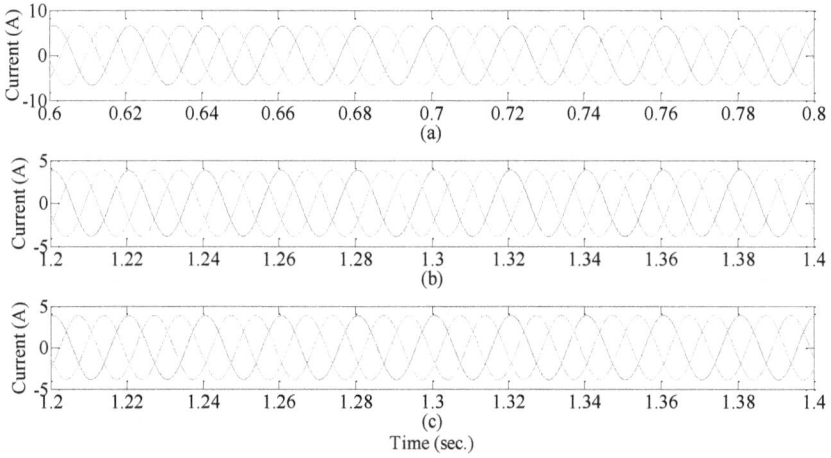

Fig. 5. Steady-state current of (a) generator 1 during breaker OFF position (b) generator 1 during breaker ON position (c) generator 2 during breaker ON position

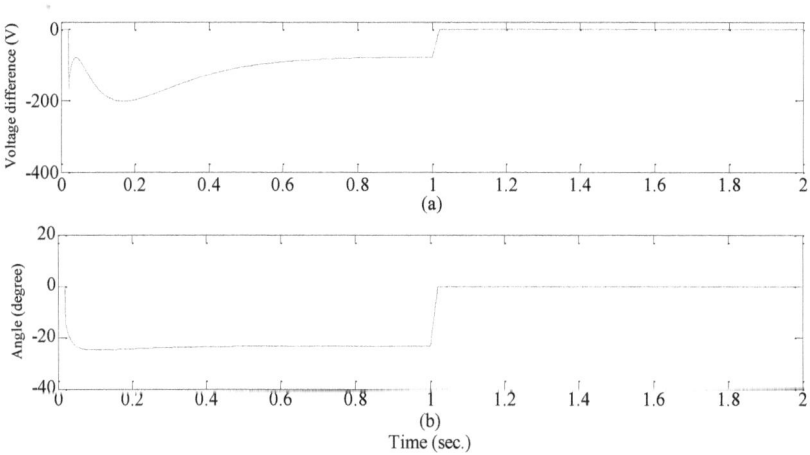

Fig. 6. Difference in (a) terminal voltage and (b) phase angle displacement of generators

Table I: Parameter of synchronous machine

S. No.	Parameter	Value
1.	Stator resistance R_s	1.62 Ω
2.	Stator leakage inductance L_l	4.527 mH
3.	Field resistance R_f	1.208 Ω
4.	Field leakage inductance L_{lfd}	11.32 mH
5.	Magnetizing inductance L_{md}	108.6 mH
6.	Magnetizing inductance L_{mq}	51.75 mH
7.	Damper resistance R_{kd}	3.142 Ω
8.	Damper Leakage inductance L_{lkd}	7.3344 mH
9.	Damper resistance R_{kq}	4.772 Ω
10.	Damper Leakage inductance L_{lkq}	10.15 mH

EXPERIMENT 9

9.1 Objective: To plot 'V' and 'inserted V' curves of a synchronous motor.

9.2 Software Required: MATLAB /Simulink software.

9.3 Brief Theory

The armature current drawn by a synchronous machine for a given power output is a function of its field current. For a given load on the machine, as the field excitation is varied, input current and power factor both change. The plot of input armature current as a function of field excitation (current) is called V curve, because of its characteristic shape. The points on the V curves, where the input armature current is minimum corresponds to unity power. The curve joining the minimum current points for different output power /load is termed as unity power factor compounding curve. As the load is increased, the range of field current for which machine gives stable operation gradually decreases.

This experiment is based on the operation of a three phase synchronous motor, considering the effect of change of field excitation at a constant load with following conditions:

i. If the armature current, I_a taken from input supply is at unity power factor, the excitation current, I_f fed to field winding is known as *normal excitation* current. In this condition, magnitude of armature current is minimum at a particular load and motor operates at unity power factor

ii. If excitation is increased above the normal excitation, the motor is said to be operated in *over excited* condition. Increasing the excitation I_f increases the armature current I_a and decreases the power factor. In over excited condition, motor operates at leading power factor.

iii. If excitation is decreased below the normal excitation, the motor is said to be operated in *under excited* condition. Decreasing the excitation I_f also increases the armature current I_a and decreases the power factor. In under excited condition, motor operates at lagging power factor.

If the armature current I_a is plotted against the field current I_f of a synchronous motor at constant load, 'V' curve is obtained. The current drawn by the motor will be minimum, when the current I_a is in phase with the input voltage or the power factor of the motor is unity. Increasing or decreasing the field current I_f from normal excitation controls the reactive power supplied to/from input supply (i.e. utility grid). Additionally, a family of curve may be drawn between the motor power factor and field current. These curves are known as 'inverted V' curve. The highest point of the curve indicates the unity power factor. More details of 'V' and 'inverted V' curves are discussed in following sections.

A general connection diagram of a three phase synchronous motor used for experimentation purpose is shown in Fig. 1. In order to vary field excitation, a rheostat is connected in rotor circuit. Input power is measured by using two-wattmeter (P_1 and P_2) and operating power factor ($\cos \emptyset$) is evaluated from the expression given below:

$$\cos \emptyset = \cos \left(\tan^{-1} \left(\frac{\sqrt{3}(P_1 - P_2)}{(P_1 + P_2)} \right) \right)$$

9.4 Experimental Procedure using MATLAB /Simulink

In order to simulate the motor operation, three-phase, 8.1 kVA, 4-poles synchronous machines is used with their parameters mentioned in Table I. Chronological steps to be followed are as given below:

Step 1: Launch Simulink in Matlab.
Step 2: Create a new model file, in which all the required blocks (synchronous machine, three-phase breaker, load etc.) are taken from "SimPowerSystem" library.
Step 3: All the blocks in model file are connected in view of Fig. 1. Developed model of complete system is shown in Fig. 2.
Step 4: In this developed system, rated three phase line voltage (400 V, 50 Hz.) is fed to armature winding. Rotor field circuit was excited with DC source V_f, and mechanical output power P_m is obtained (power fed with negative sign in the model). Suitable parameters are fed into the blocks through "Block parameters" dialogue box. For ready reference, it is shown, for three-phase synchronous machine and power supply, in Fig. 3 (a) and Fig. 3 (b) respectively.
Step 5: Select the suitable solver type and run the developed model. In present simulation, "Fixed-step" with 0.001 sec. as step size, ode 4

(Runge-Kutta) solver type was selected. Developed model is run at a particular load at different field excitation.

Step 6: Responses of motor are recorded.

Step 7: Plot the 'V' and 'inverted V' curves.

During the simulation of motor operation, readings were recorded for two operating loads (i.e. no-load and at mechanical load P_m = 1kW). Reading of the recorded data during experiment is given in observation Table II (a) and (b). From the recorded readings, 'V' curve and 'inverted V' curves are plotted in Fig. 4 and Fig. 5 respectively for motor operation at no-load and 1 kW load.

Readers are advised to simulate the developed system different loads for each set of reading in the format shown in Table III.

9.5 Results and Discussion

Test of a three-phase synchronous motor was conducted by using MATLAB/Simulink observations were noted (armature current, field current and power factor) and both 'V' and 'inserted V' curves were plotted.

In case of synchronous motor, for a given load, if the field excitation is increased beyond the field current corresponding to unity power factor, armature current increases and the motor operates at leading power factor in over-excited condition. If the field current is decreased below the field current corresponding to unity power factor, motor operates at lagging power factor in under-excited condition. The case is completely reversed in case of synchronous alternator. In both the cases of synchronous alternator and motor, with the increase in load, V curve shift upward, and stable operational zone goes on decreasing.

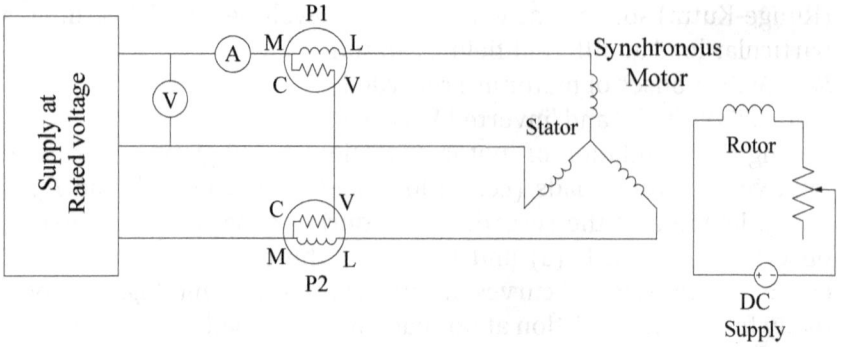

Fig. 1. Connection diagram of a three phase synchronous motor

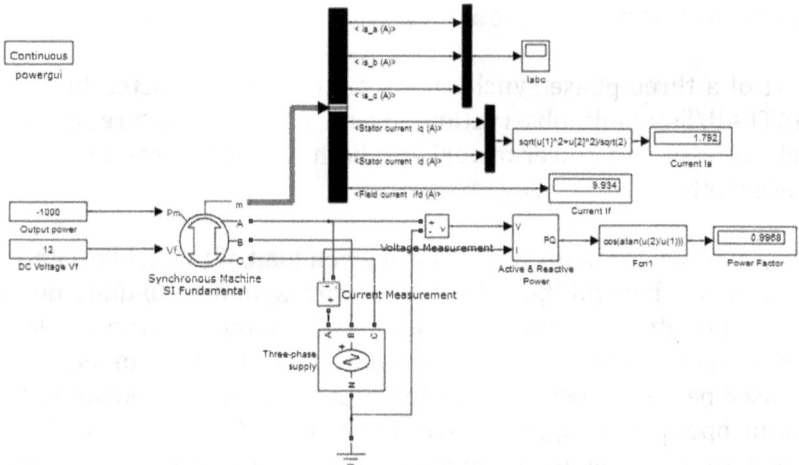

Fig. 2. Simulink model of synchronous motor operation

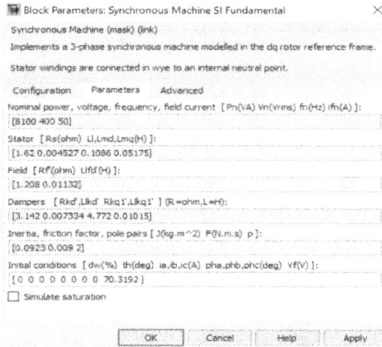
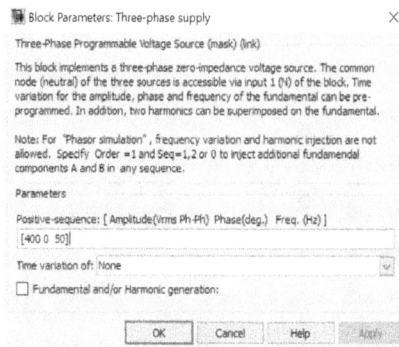

(a)   (b)

Fig. 3. "Block parameters" dialog box (a) three-phase synchronous machine (b) three-phase supply

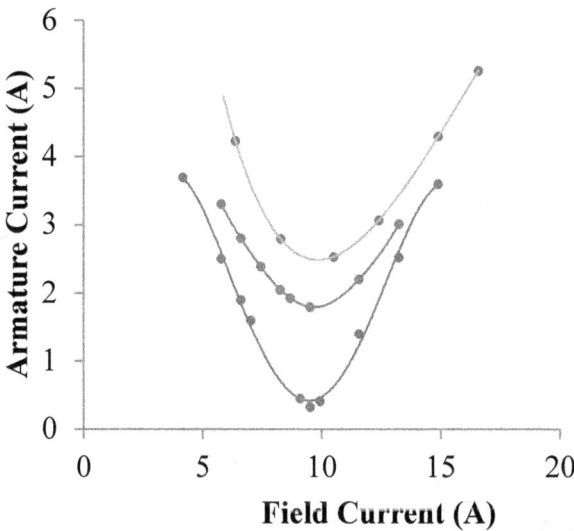

Fig. 4. 'V' curve plot of motor at (a) no-load (b) 1kW load

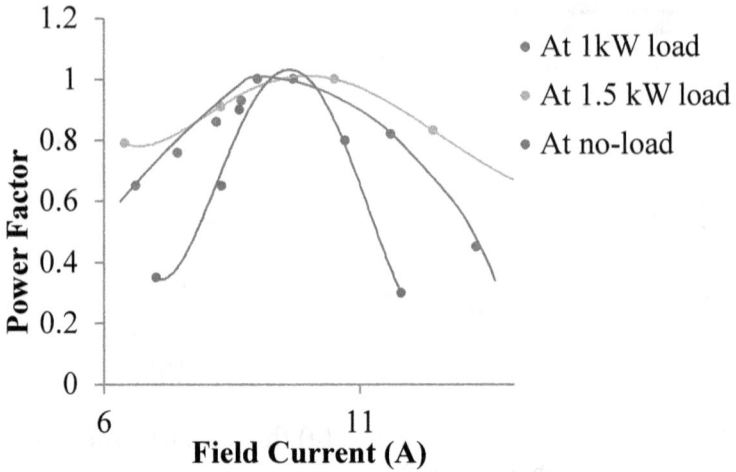

Fig. 5. 'Inverted V' curve of motor at (a) no-load (b) 1kW load

Table I: Parameter of synchronous machine

S. No.	Parameter	Value
1.	Stator resistance R_s	1.62 Ω
2.	Stator leakage inductance L_l	4.527 mH
3.	Field resistance R_f	1.208 Ω
4.	Field leakage inductance L_{lfd}	11.32 mH
5.	Magnetizing inductance L_{md}	108.6 mH
6.	Magnetizing inductance L_{mq}	51.75 mH
7.	Damper resistance R_{kd}	3.142 Ω
8.	Damper Leakage inductance L_{lkd}	7.3344 mH
9.	Damper resistance R_{kq}	4.772 Ω
10.	Damper Leakage inductance L_{lkq}	10.15 mH

Table II (a): Observation at no-load

S. No.	Field voltage V_f (V)	Field Current I_f (A)	Armature Current I_a (A)	Power Factor	
1	5.0	4.14	3.70	0.11	Lagging
2	7.0	5.80	2.60	0.14	
3	8.5	7.04	1.74	0.20	
4	11.5	9.52	0.32	1.00	Unity
5	12.0	9.93	0.41	0.79	Leading
6	14.0	11.59	1.42	0.24	
7	16.0	13.25	2.52	0.15	
8	18.0	14.9	3.64	0.11	

Table II (b): Observation at 1kW load

S. No.	Field voltage V_f (V)	Field Current I_f (A)	Armature Current I_a (A)	Power Factor	
1	7.0	5.80	3.27	0.56	Lagging
2	8.0	6.62	2.80	0.65	
3	9.0	7.45	2.38	0.76	
4	10.0	8.28	2.05	0.88	
5	10.5	8.70	1.94	0.93	
6	11.5	9.52	1.79	1.00	Unity
7	12.0	9.93	1.80	0.99	Leading
8	14.0	11.59	2.20	0.82	
9	16.0	13.25	3.01	0.61	

Table III: Observation format at a load

S. No.	Field voltage V_f (V)	Field Current I_f (A)	Armature Current I_a (A)	Power Factor	
1					
2					
3					
4					
5					
6					
7					

EXPERIMENT 10

10.1 Objective: To perform closed loop operation of three-phase induction motor using vector control.

10.2 Software Required: MATLAB /Simulink software.

10.3 Brief Theory:

'Vector Controllers' name is given, because these controllers control both the amplitude and phase of the ac excitation. The vector control of current and voltages results in control of the spatial orientation of the electromagnetic fields in the machine and has led to the term 'Field Orientation'. A vector control (i.e. field oriented control) scheme is adopted to operate the induction motor similar to that of a DC separately exited motor, wherein an independent control of flux and torque component of current is achieved. For this purpose, motor is operated in synchronously rotating reference frame $(d^e - q^e)$, where the sinusoidal ac variables become dc quantities under steady-state.

In this experiment, mathematically developed dynamic model in two axes $(d - q$ axis) in arbitrary reference frame is used in control scheme. Axis transformation with a mathematical modeling of induction motor using indirect field oriented control has been reviewed below:

10.3.1 Axis Transformation

A two axis transformation of a symmetrical three-phase system is shown in Fig. 1. Considering the angular displacement of two-phase axis $(d^s - q^s)$ by θ with respect to three-phase $(a - b - c)$ axis, transformed voltages are be expressed as

$$\begin{bmatrix} v_q^s \\ v_d^s \\ v_0^s \end{bmatrix} = \frac{2}{3} \begin{bmatrix} \cos\theta & \cos(\theta - 2\pi/3) & \cos(\theta + 2\pi/3) \\ \sin\theta & \sin(\theta - 2\pi/3) & \sin(\theta + 2\pi/3) \\ 0.5 & 0.5 & 0.5 \end{bmatrix} \begin{bmatrix} v_a \\ v_b \\ v_c \end{bmatrix} \qquad (10.1)$$

In a balanced system, zero sequence component v_0^s is neglected. For convenience, q^s axis is assumed to be coinciding with a axis. Hence for $\theta = 0$, voltage equation in matrix form is simplified as

$$\begin{bmatrix} v_q^s \\ v_d^s \end{bmatrix} = \begin{bmatrix} 1 & 0 & 0 \\ 0 & -1/\sqrt{3} & 1/\sqrt{3} \end{bmatrix} \begin{bmatrix} v_a \\ v_b \\ v_c \end{bmatrix} \qquad (10.2)$$

Two axis $(d^s - q^s)$ in stationary reference frame is transformed to synchronously rotating reference frame (i.e. $d^e - q^e$ axis) having the angular displacement of θ_e $(= \omega_e t)$ is shown in Fig. 2. Hence, quantities in $d^s - q^s$ axes may be transformed to $d^e - q^e$ axis with following matrix representation:

$$\begin{bmatrix} v_q^e \\ v_d^e \end{bmatrix} = \begin{bmatrix} \cos \theta_e & -\sin \theta_e \\ \sin \theta_e & \cos \theta_e \end{bmatrix} \begin{bmatrix} v_q^s \\ v_d^s \end{bmatrix} \qquad (10.3)$$

By taking the inverse of above transformation will yield the transformed quantities in $d^s - q^s$ axis from $d^e - q^e$ axis.

$$\begin{bmatrix} v_q^s \\ v_d^s \end{bmatrix} = \begin{bmatrix} \cos \theta_e & \sin \theta_e \\ -\sin \theta_e & \cos \theta_e \end{bmatrix} \begin{bmatrix} v_q^e \\ v_d^e \end{bmatrix} \qquad (10.4)$$

10.3.2 Mathematical Modeling of Induction motor drives

The voltage equation of stator and rotor circuit in arbitrary reference frame is given as

$$v_{qs} = r_s i_{qs} + \frac{\omega_r}{\omega_b} \psi_{ds} + \frac{p}{\omega_b} \psi_{qs} \qquad (10.5)$$

$$v_{ds} = r_s i_{ds} - \frac{\omega_r}{\omega_b} \psi_{ds} + \frac{p}{\omega_b} \psi_{ds} \qquad (10.6)$$

$$v_{qr} = r_r i_{qr} + \left(\frac{\omega - \omega_r}{\omega_b}\right) \psi_{dr} + \frac{p}{\omega_b} \psi_{qr} \qquad (10.7)$$

$$v_{dr} = r_r i_{dr} - \left(\frac{\omega - \omega_r}{\omega_b}\right) \psi_{qr} + \frac{p}{\omega_b} \psi_{dr} \qquad (10.8)$$

In the above expression, flux linkage per second is determined as

$$\psi_{qs} = x_{ls} i_{qs} + x_m \left(i_{qs} + i_{qr}\right) \qquad (10.9)$$

$$\psi_{ds} = x_{ls} i_{ds} + x_m \left(i_{ds} + i_{dr}\right) \qquad (10.10)$$

$$\psi_{qr} = x_{lr}i_{qr} + x_m(i_{qs} + i_{qr}) \tag{10.11}$$
$$\psi_{dr} = x_{lr}i_{dr} + x_m(i_{ds} + i_{dr}) \tag{10.12}$$

Subscript 's' and 'r' signifies stator and rotor circuit respectively. Developed motor torque and rotor speed is given by

$$T_e = \left(\frac{3}{2}\right)\left(\frac{P}{2}\right)\left(\frac{1}{\omega_b}\right)x_m(i_{qs}i_{dr} - i_{ds}i_{qr}) \tag{10.13}$$

$$\omega_r = \frac{P}{2J}\int(T_e - T_l)dt \tag{10.14}$$

The rotor equation in rotating reference frame ($\omega = \omega_e$) may be rewritten as

$$0 = r_r i_{qr} + \frac{\omega_{sl}}{\omega_b}\psi_{dr} + \frac{p}{\omega_b}\psi_{qr} \tag{10.15}$$

$$0 = r_r i_{dr} - \frac{\omega_{sl}}{\omega_b}\psi_{qr} + \frac{p}{\omega_b}\psi_{dr} \tag{10.16}$$

where

$$\omega_{sl} = \omega_e - \omega_r \tag{10.17}$$

and superscript 'e' has been dropped in following section for easier representation.

Assuming the rotor flux is aligned only along d axis, therefore

$$\psi_r = \psi_{dr} \tag{10.18}$$

$$\psi_{qr} = 0 \tag{10.19}$$

$$p\omega_r = 0 \tag{10.20}$$

Substitution of equations (10.18) – (10.20) into (10.15) and (10.16) yields,

$$r_r i_r + \frac{\omega_{sl}}{\omega_b}\psi_r = 0 \tag{10.21}$$

$$r_r i_r + \frac{p}{\omega_b}\psi_{dr} = 0 \tag{10.22}$$

90

The value of i_{qr} and i_{dr} obtained from equation (10.11) – (10.12), substituted in equation (10.21) – (10.22) yield

$$\omega_{sl} = K_{sl}i_T \qquad (10.19)$$

$$i_F = (1 + T_r p)\psi_r \qquad (10.20)$$

where

$$K_{sl} = \frac{X_m}{T_r \omega_b \psi_r} \qquad (10.20\ A)$$

$$i_T = i_{qs} \qquad (10.20\ B)$$

$$i_F = i_{ds} \qquad (10.20\ C)$$

$$T_r = \frac{L_r}{r_r} \text{ is rotor time constant} \qquad (10.20\ D)$$

The value of rotor current when substituted in torque equation will yield the following after simplification

$$T_e = K_T \psi_r i_T = K_1 i_T \qquad (10.21)$$

$$K_1 = K_T \psi_r \qquad (10.21\ A)$$

$$K_T = c \frac{L_m}{L_r} \qquad (10.21\ B)$$

$$c = \left(\frac{3}{2}\right)\left(\frac{P}{2}\right)\left(\frac{1}{\omega_b}\right) \qquad (10.21\ C)$$

Implementation of the closed loop operation using indirect field oriented control is shown in Fig. 3. Initially, rotor speed (ω_r) is sensed and compared with its reference value (ω_r^*). Speed error signal is then fed to a proportional-integrator (PI) controller and torque component of current (i.e. active component of current i_{qs}^*) is generated. Reference value of flux component of current (i.e. reactive component of current i_{ds}^*) is kept constant, because motor operation in field weakening region is not considered in this experiment. Hence, the reference current (i_{qs}^* and i_{ds}^*) is constant in synchronous

91

reference frame, and are transformed to three axes $(a - b - c)$ in stationary reference frame in following two steps:

i) Transformation of reference current $(i_{qs}^*$ and $i_{ds}^*)$ to two axis stationary reference frame by using equation 4.

ii) Transformation of two-phase stationary current $(i_{qs}$ and $i_{ds})$ to three-phase current $(i_a, i_b$ and $i_c)$.

Hence, obtained three-phase reference current $(i_a^*, i_b^*$ and $i_c^*)$ is compared with actual motor current $(i_a, i_b$ and $i_c)$, and error current is fed to hysteresis controller, wherein suitable current band is set to generate the switching signal to the power electronics switches of inverter circuit. The expressions of phase voltages in a three-phase inverter using switching functions are as follows:

$$v_a = \left(\frac{V_{dc}}{3}\right)[2S_a - S_b - S_c] \tag{10.22}$$

$$v_b = \left(\frac{V_{dc}}{3}\right)[2S_b - S_a - S_c] \tag{10.23}$$

$$v_c = \left(\frac{V_{dc}}{3}\right)[2S_c - S_b - S_a] \tag{10.24}$$

where
S_a, S_b, S_c are the switching function associated with phase a, b and c respectively. V_{dc} is the dc-link voltage.

10.3 Experimental Procedure using MATLAB /Simulink

Step 1: Launch Simulink in Matlab.
Step 2: Create a new model file, in which all the required blocks are taken from "Simulink" library.
Step 3: All the required components of drive system is made through their mathematical modeling. Complete motor drive system developed for operation in closed loop is shown in Fig. 4. Developed subsystem Simulink model of induction motor, reference current generation using vector control, inverter circuit and transformation used are shown in Appendix, Fig. A1, Fig. A2, Fig. A3 and Fig. A4, respectively.

Step 4: Select the suitable solver type and run the developed model. In present simulation, "Fixed-step" with 0.0001 sec. as step size, ode 4 (Runge-Kutta) solver type was selected.
Step 5: Record and display the motor drive response.

In this experiment, 4 poles, 2 HP, three-phase, induction machine are considered with the parameter given in Table I. Initially, sensed motor speed is compared from its reference command at 157.08 rad/sec to generate torque component of current $i_T^* = i_{qs}^*$ through a PI controller. Reference value of flux component of current $i_F^* = i_{ds}^*$) is set at 1.5 A. Generated reference current in synchronous frame is transformed to stationary frame by using equation 4, in which angular position is θ_e $(= \theta_r + \theta_{sl})$. Transformed generated reference current in stationary three-phase current $(i_a^*, i_b^*$ and $i_c^*)$ is obtained by using equation 2. Actual motor current $(i_a, i_b$ and $i_c)$ are compared with generated reference current $(i_a^*, i_b^*$ and $i_c^*)$ in comparator circuit. Error current in each phase is fed to the hysteresis controller, wherein lower and upper current limit is set at -0.025 A and 0.025 A respectively. Hence, the switching signals generated, S_a, S_b and S_c at the output of hysteresis controller are used to operate the inverter circuit, which is connected to a constant DC link voltage at 600 V.

During machine operation, speed command of 157.08 rad/sec. was initiated at time $t = 0.2$ sec., which increases at a slope of 483.32 rad/sec² by using rate limiter block, and motor continue to operate at no-load. At time $t = 2$ sec., a load of 8.0 Nm. is applied, resulting in a small speed dip by 1.28 rad/sec, but within 0.2 sec. motor regains the commanded speed. Zoomed portion of the speed response showing a dip in speed is shown in Fig. 5 (a). At time $t = 3$ sec., speed reversal command was initiated and at $t = 3.8$ sec., motor starts to operate in reversed motoring mode under steady-state. Developed torque response of the motor is shown in Fig. 5 (b). Application of load results in the increase in motor current during motor and reverse motoring mode under steady-state. Motor input phase current response of armature winding is shown in Fig. 6.

10.5 Results and Discussion

Three-phase induction motor was operated in closed loop using vector control on virtual platform of MATLAB/Simulink. Four quadrant

93

operation was simulated by using the developed drive system. During the machine operation, an improved operation was noted due to an independent control of torque and field component of stator current.

One can use either current regulated Voltage Source Inverter (VSI) or a Current Source Inverter (CSI). In CSI, a large inductor is used at the dc side, thus limiting the available rate of change of current. It is for this reason that a standard solution for vector controlled drive utilizes the current regulated VSI.

Hysteresis current control is susceptible to noise and not suitable for higher power ratings that requires a relatively low switching frequency. Predictive current controller increases the computational burden of the digital signal controller. Synchronous reference (d-q) control requires back and forth transformation with several trigonometric calculations and additional sophistication to the controller with a corresponding intricate tuning process. Space Vector Modulation Technique (SVM) utilizes dc bus voltage source more efficiently and generates less harmonic distortions as compared to sinusoidal (SPWM) technique.

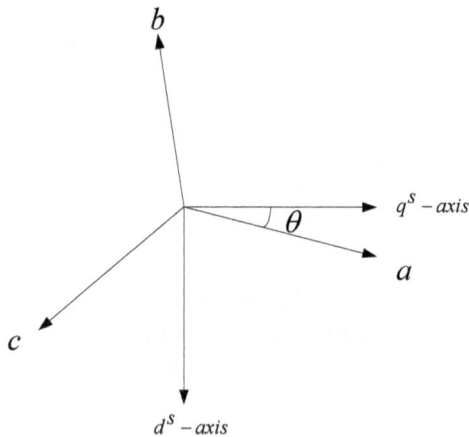

Fig. 1. Three-phase axes transformation to two-phase in stationary frame

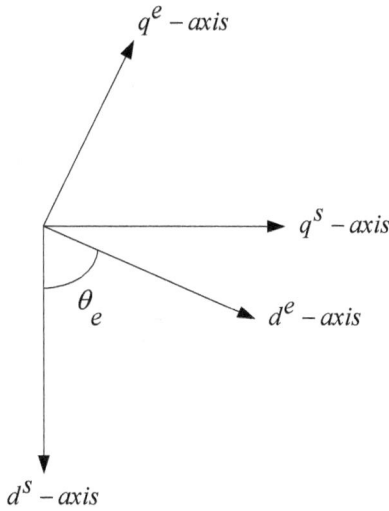

Fig. 2. Two axes transformation from stationary to rotating reference frame

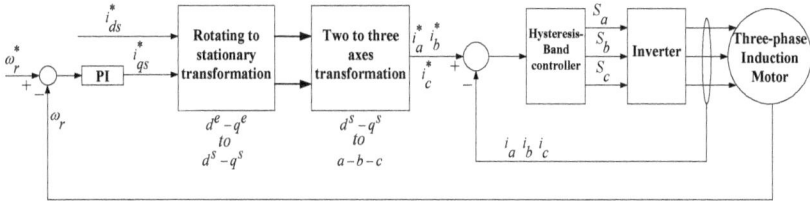

Fig. 3. Implementation of induction motor drive

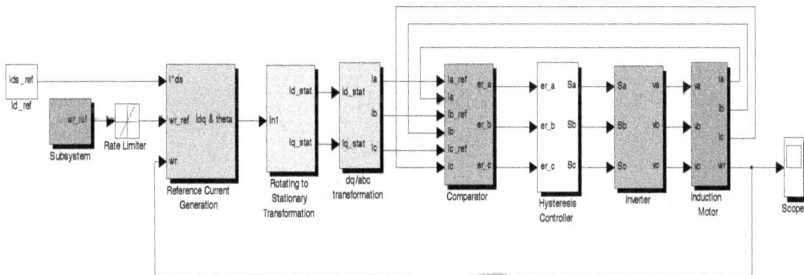

Fig. 4. Complete developed model of indirect field oriented controlled motor drive

95

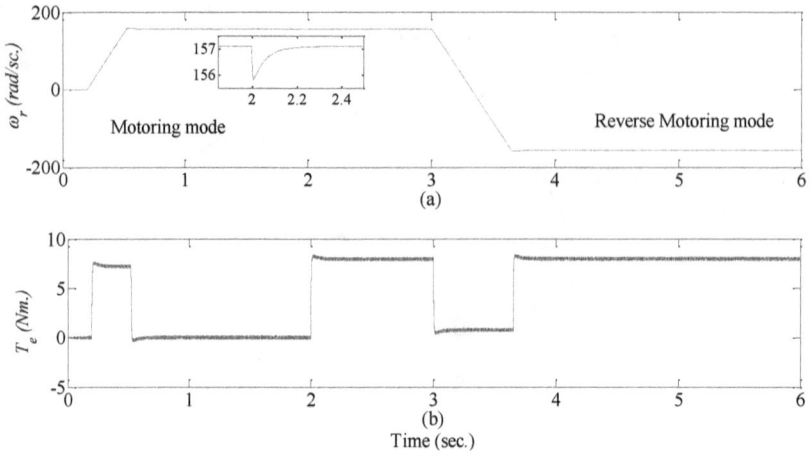

Fig. 5. Motor response (a) Rotor speed (b) Torque developed

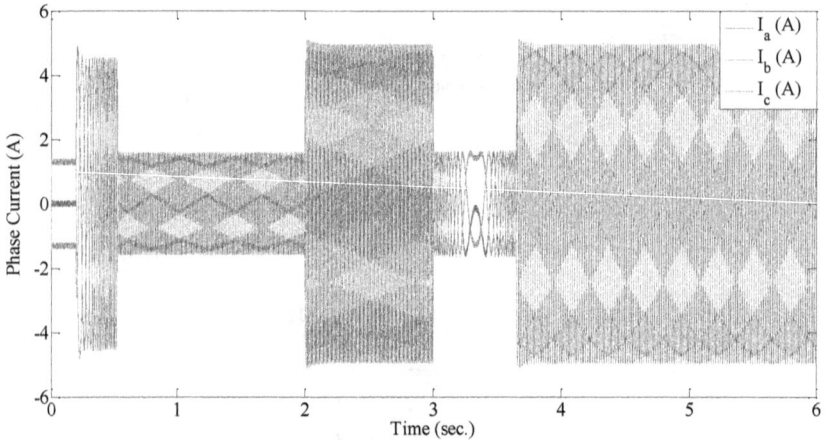

Fig. 6. Armature current of induction motor

Table I: Parameters of 4 poles, 2 HP, three-phase induction machine

$r_s = 10\ \Omega$	$r_r = 6.3\ \Omega$
$L_{ls} = L_{lr} = 0.04\ H$	$L_m = 0.42\ H$
	$J = 0.03\ Kgm^2$

Appendix

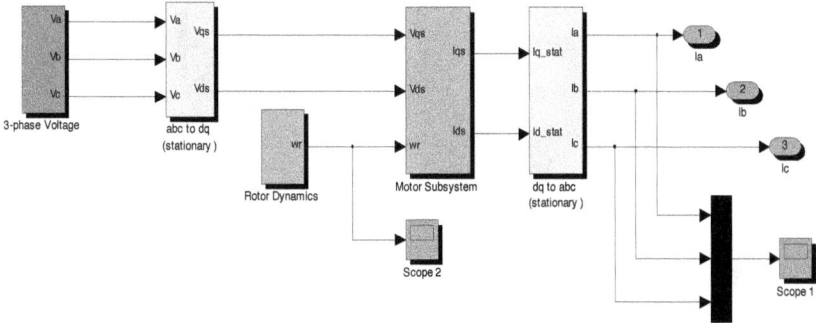

Fig. A1. Developed Simulink model of three-phase induction motor

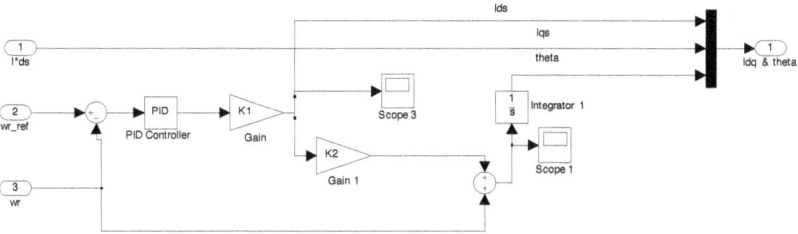

Fig. A2. Reference current generation in synchronous frame

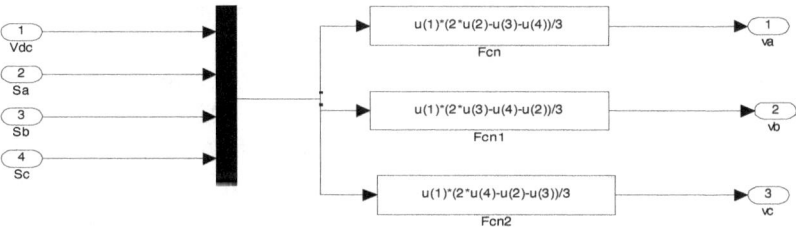

Fig. A3. Model of three-phase inverter

(a)

(b)

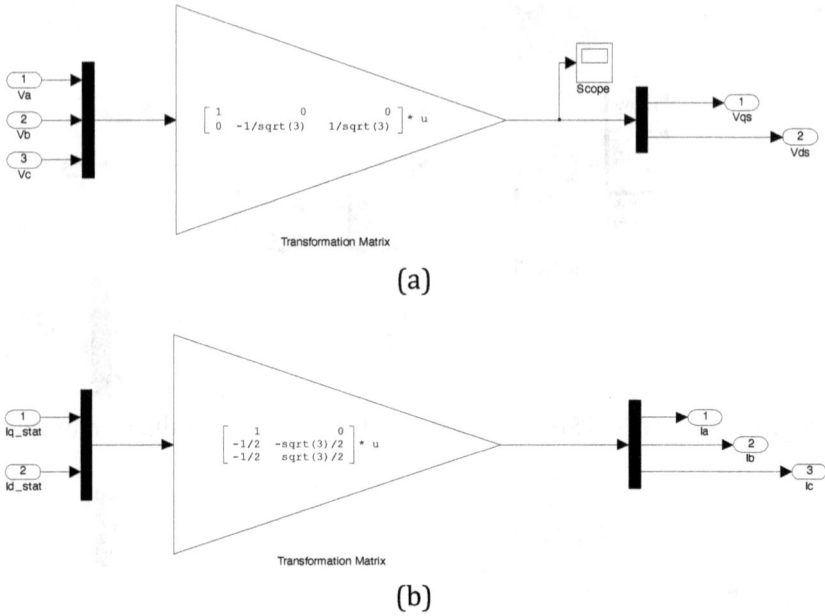

Fig. A4. Subsystem showing the transformation from (a) *abc* to *dq* and (b) *dq* to *abc* in stationary frame

References

1. S. J. Chapman, "Electric Machinery Fundamentals", McGraw Hill, 4th edition, 2005.
2. R. Krishnan, "Electric Motor Drives Modeling, Analysis, and Control", Pearson India Education, Inc., 2015.
3. P. C. Krause, O. Wasynczuk, S. D. Sudhoff, "Analysis of electrical machinery and drive systems", IEEE Press, 2nd edition, 2002.
4. C. M. Ong, "Dynamic Simulation of Electrical Machinery using Matlab/Simulink", Prentice Hall PTR, 1998.
5. A. Husain, "Electric Machines", Dhanpat Rai & Co. (Pvt.) Ltd., New Delhi, India, 3rd edition, 2016.

Part B

Experiments with MULTISIM

EXPERIMENT 1

1.1 Objective: To verify superposition theorem of a network.

1.2 Software Used: MULTISIM or similar software.

1.3 Brief Theory:

Superposition theorem is a general principle, which enables to determine the effect of a number of available energy sources (voltage and/or current source) simultaneously in a network by considering the effect of each source individually and combining (i.e. superimposing) their effects. Statement of superposition theorem is state as:

In a linear network, having more than one independent sources (voltage and current source), overall response (in terms of current and/or voltage) in any branch is obtained by the summation of response due to each independent source connected one at a time with other sources equated to zero. Hence, during the evaluation of response by considering a particular source, other current sources are equated to zero by replacing it to an open-circuit whereas, voltage sources are equated to zero by replacing it to a short-circuit. Superposition theorem is very helpful in the network excited by with a combination dc and ac sources or ac sources with different frequencies.

1.4 Experiment Procedure using MULTISIM

Step 1: Open MULTISIM and create a new design file to develop the circuit.
Step 2: Make the connections as shown in the circuit diagram in Fig. 1 by using MULTISIM. Circuit connection using MULTISIM is shown in Fig. 2. In this circuit, load current was $I_L = 0.111\ A$ with the direction shown by arrow in figure.
Step 3: Considering only 10 V, other voltage sources (15 V and 8 V) are shorted, load current was $I_{L1} = 0.267\ A$, as shown in Fig. 3.
Step 4: Considering only 15 V, other voltage sources (10 V and 8 V) are shorted, load current was $I_{L2} = -0.333\ A$, as shown in Fig. 4. Negative sign signifies the direction of load current is opposite to assumed direction.

Step 5: Considering only 8 V, other voltage sources (10 V and 15 V) are shorted, load current was $I_{L3} = 0.178\ A$, as shown in Fig. 5. Load current in above steps are noted in Table I.

Table I: Observation of load current using superposition theorem

Current I_{L1} (A) with 10 V suppy	Current I_{L2} (A) with 15 V suppy	Current I_{L2} (A) with 8 V suppy	Total current (I_L) with all voltage sources
0.267	−0.333	0.178	0.111

Hence, total load current is obtained by the summation of currents I_{L1}, I_{L2} and I_{L3}

$$I_L = I_{L1} + I_{L2} + I_{L3}$$
$$= 0.267 - 0.333 + 0.178$$
$$= 0.111\ A$$

Obtained total load current is same when all the sources are acting simultaneously in the circuit shown in Fig. 2. Hence, superposition theorem is verified.

1.5 Result:

Superposition theorem of the given circuit is verified using MULTI-SIM.

Readers are suggested to develop MULTISIM model for other networks and follow the above procedural steps to verify the Superposition theorem.

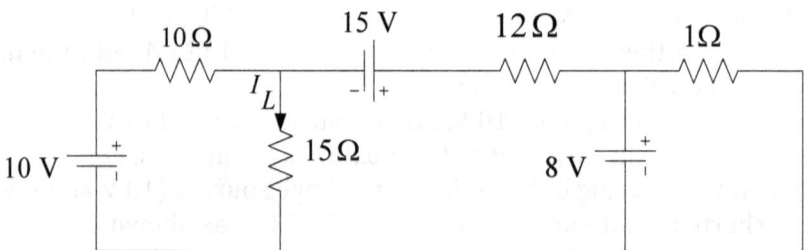

Fig. 1. Circuit diagram

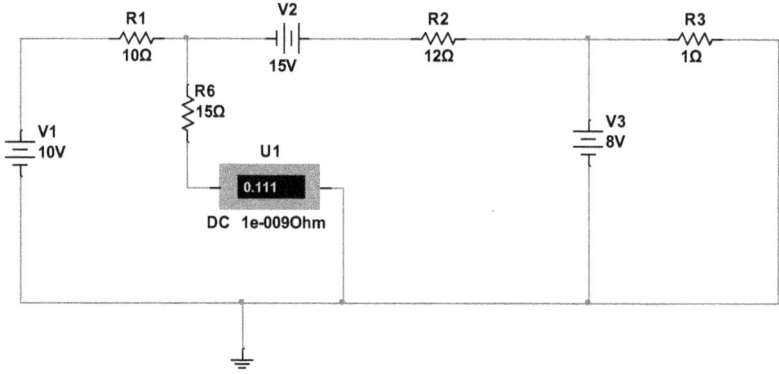

Fig. 2. Circuit connection in MULTISIM

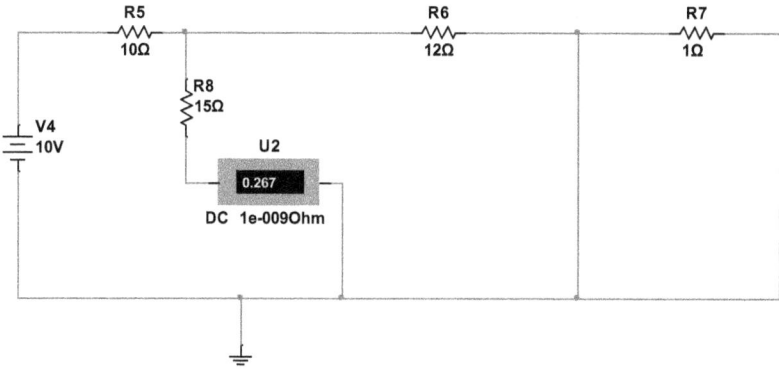

Fig. 3. Circuit with only 10 V supply

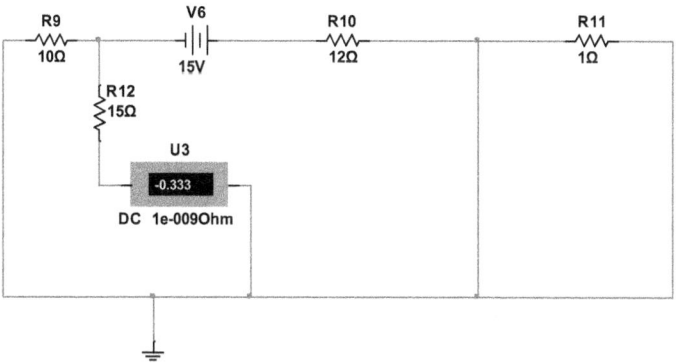

Fig. 4. Circuit with only 15 V supply

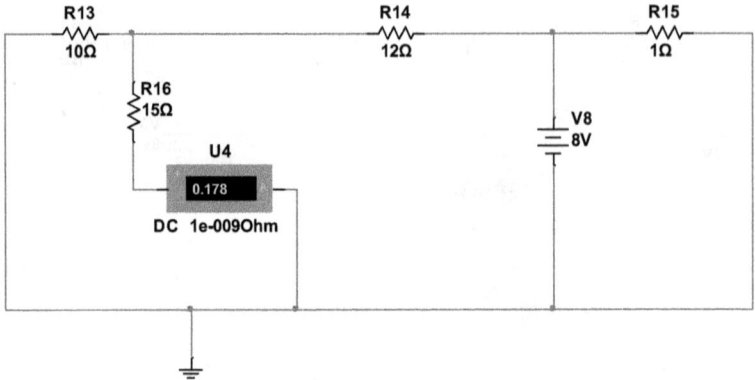

Fig. 5. Circuit with only 8 V supply

EXPERIMENT 2

2.1 Objective: To verify Thevenin's theorems of a network.

2.2 Software Used: MULTISIM or similar software.

2.3 Brief Theory:

During network analysis, sometimes it is desirable to determine the response (i.e. voltage and/or current) in only one element. So, it is not necessary to analyse the complete circuit and response is evaluated by using Thevenin's theorem. Statement of Thevenin's theorem is state as:

A linear two-terminal active network consisting of N voltage and current sources (independent and/or dependent) can be replaced by an equivalent voltage source connected in series with impedance. An active DC network depicted in Fig. 1 (a) is shown with its Thevenin's equivalent circuit in Fig. 1 (b), wherein V_{Th} and R_{Th} is known as Thevenin's equivalent voltage and resistance respectively. Following steps are adopted to determine the Thevenin's parameters (voltage V_{Th} and resistance R_{Th}) across a load resistance R_L:

1. Load branch with resistance R_L across which response (i.e. current) are to be evaluated is removed from the circuit. Hence open-circuit voltage V_{oc}, across open-ended terminals a-b gives Thevenin's voltage V_{Th} $(= V_{oc})$.
2. Opened terminals a-b are shorted and short-circuit current I_{sc} is noted.
3. Thevenin's resistance R_{Th} is determined by using following relation:
$$R_{Th} = \frac{V_{oc}}{I_{sc}} = \frac{V_{Th}}{I_{sc}}$$
$$(2.1)$$
4. Hence Thevenin's parameters (voltage V_{Th} and resistance R_{Th}) found in above steps are used to draw the Thevenin's equivalent circuit by connected voltage source V_{Th} in series with resistance R_{Th}.

2.4 Experiment Procedure using MULTISIM

Step 1: Open MULTISIM and create a new design file to develop the circuit.

Step 2: Make the connections as shown in the circuit diagram in Fig. 2 by using MULTISIM. Circuit connection using MULTISIM is shown in Fig. 3. In this circuit, load current was $I_L = 0.111\ A$ with the direction shown by arrow in Fig. 2.

Step 3: Remove the load resistance R_L (15 Ω) from the considered circuit, creating an opened terminals. Note down the open-circuit voltage V_{oc} to obtain Thevenin's voltage V_{Th} by connecting a voltmeter as shown in Fig. 4.

Step 4: Short circuit the opened terminals of step 3. Note down the short circuit current I_{sc} by connecting an ammeter as shown in Fig. 5.

Step 5: Calculate Thevenin's resistance R_{Th}. All the readings are noted in observation Table I.

Step 6: Design the Thevenin's equivalent circuit and connect the load resistance R_L. Note down the load current as shown in Fig. 6. Value of load current was found to be equal in both direct measurement (in Fig. 3) and Thevenin's equivalent circuit. Hence, Thevenin's theorem is verified.

2.5 Result

Thevenin's theorem of the given circuit is verified using MULTISIM.

Readers may vary one or more voltage source and determine Thevenin's equivalent parameters (V_{Th} and R_{Th}). In all the case, value of resistance R_{Th} will be same. Further, it is suggested to develop MULTISIM model for other networks and follow the above procedural steps to verify the Thevenin's theorem.

Table I: Observation of load current using Thevenin's theorem

Using Thevenin's theorem				Direct Measurement
Open circuit voltage V_{oc} (V)	Short circuit current I_{sc} (A)	Thevenin's resistance R_{Th} (Ω)	Load current I_L (A)	Load current I_L (A)
2.273	0.417	5.455	0.111	0.111

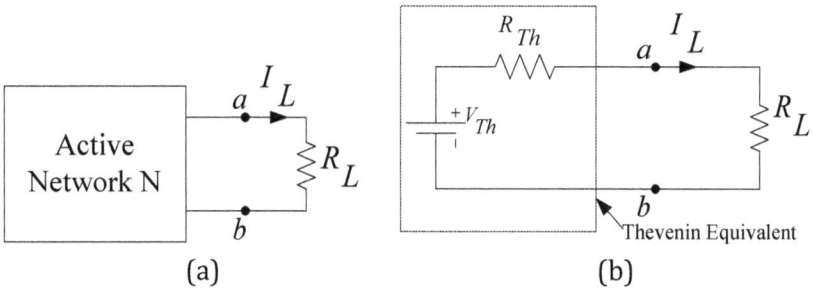

(a)　　　　　　(b)

Fig. 1. (a) Original circuit (b) Thevenin's equivalent circuit

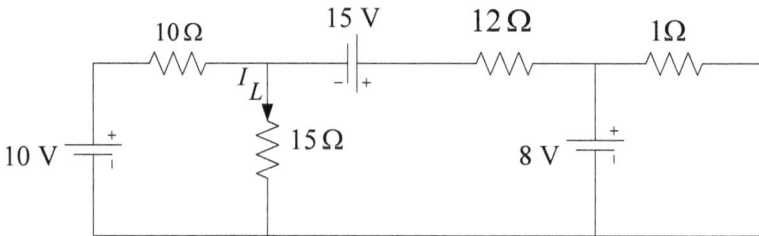

Fig. 2. Circuit diagram

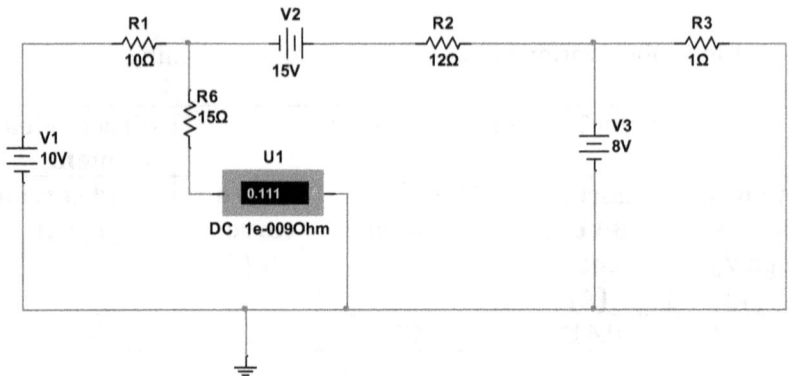

Fig. 3. Circuit connection in MULTISIM

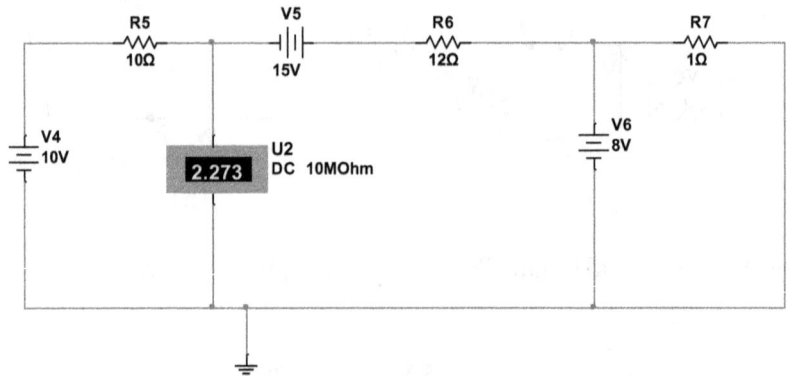

Fig. 4. Measurement of open circuit voltage

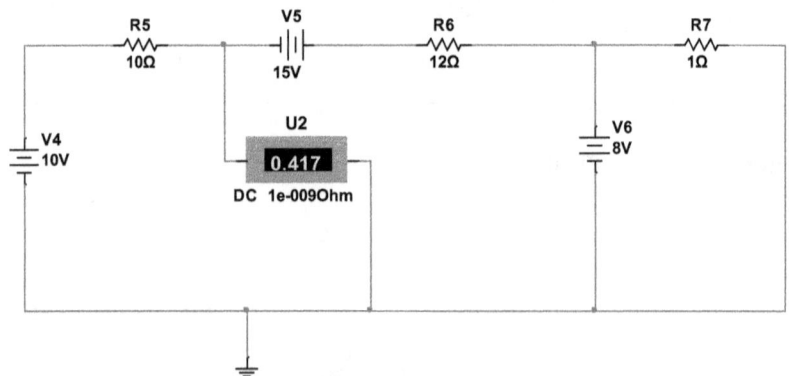

Fig. 5. Measurement of short circuit current

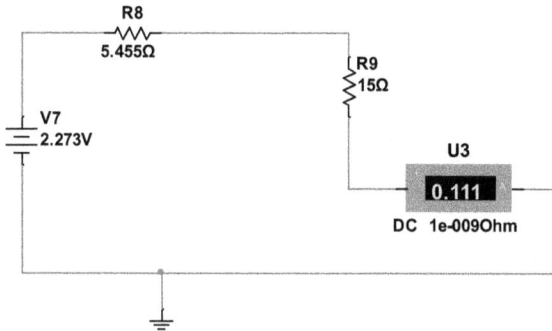

Fig. 6. Current measurement using Thevenin's equivalent circuit

EXPERIMENT 3

3.1 Objective: To verify Maximum Power Transfer theorem of a network.

3.2 Software Used: MULTISIM or similar software.

3.3 Brief Theory:

Norton's theorem is dual of the Thevenin's theorem used to determine the response (i.e. voltage and/or current) in only one element and is not necessary to analyse the complete circuit. Statement of Norton's theorem is state as:

A linear two-terminal active network consisting of N voltage and current sources (independent and/or dependent) can be replaced by an equivalent current source connected in parallel with impedance. For an active DC network in Fig. 1 (a) is shown with their Norton's equivalent circuit in Fig. 1 (b), wherein I_N and R_N is known as Norton's equivalent current and resistance respectively. Following steps are adopted to determine the Norton's parameters (current I_N and resistance R_N) across a load resistance R_L:

1. Load branch with resistance R_L across which response (i.e. current) are to be evaluated is removed from the circuit. Opened end terminals *a-b* are shorted and short-circuit current I_{sc} is noted. This current is Norton's current I_N.

2. Procedure to find internal Norton's resistance R_N is same as explained for Thenenin's resistance R_{Th}. Hence, after noting open-circuit voltage V_{oc}, across open-ended terminals *a-b*, Norton's resistance R_N is determined by using following relation:

$$R_N = \frac{V_{oc}}{I_{sc}} \tag{3.1}$$

3. Hence Norton's parameters (current I_N and resistance R_N) found in above steps are used to draw the Norton's equivalent circuit by connecting current source I_N in parallel with resistance R_N.

3.4 Experiment Procedure using MULTISIM

Step 1: Open MULTISIM and create a new design file to develop the circuit.

Step 2: Make the connections as shown in the circuit diagram in Fig. 2 by using MULTISIM. Circuit connection using MULTISIM is shown in Fig. 3. In this circuit, load current was $I_L = 0.111\ A$ with the direction shown by arrow in Fig. 2.

Step 3: Remove the load resistance R_L (15 Ω) from the considered circuit, creating an opened terminals. Note down the open-circuit voltage V_{oc} by connecting a voltmeter as shown in Fig. 4.

Step 4: Short circuit the opened terminals of step 3. Note down the short circuit current I_{sc} by connecting an ammeter to obtain Norton's current I_N as shown in Fig. 5.

Step 5: Calculate Norton's resistance R_N. All the readings are noted in observation Table I.

Step 6: Design the Norton's equivalent circuit and connect the load resistance R_L. Note down the load current as shown in Fig. 6. Value of load current was found to be equal in both direct measurement (in Fig. 3) and Norton's equivalent circuit. Hence, Norton's theorem is verified.

3.5 Results

Norton's theorem of the given circuit is verified using MULTISIM.

Readers may vary one or more voltage source and determine Norton's equivalent parameters (I_N and R_N). In all the cases, value of resistance R_N will be the same. Further, it is suggested to develop MULTISIM model for other networks and follow the above procedural steps to verify the Norton's theorem.

Table I: Observation of load current using Norton's theorem

Using Norton's theorem				Direct Measurement
Open circuit voltage V_{oc} (V)	Short circuit current I_{sc} (A)	Norton's resistance R_{Th} (Ω)	Load current I_L (A)	Load current I_L (A)
2.273	0.417	5.455	0.111	0.111

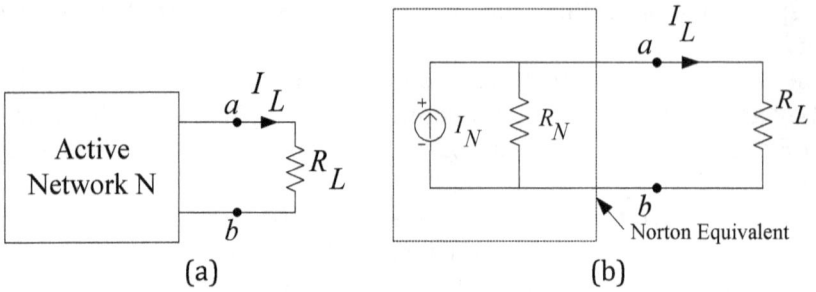

Fig. 1. (a) Original circuit (b) Norton's equivalent circuit

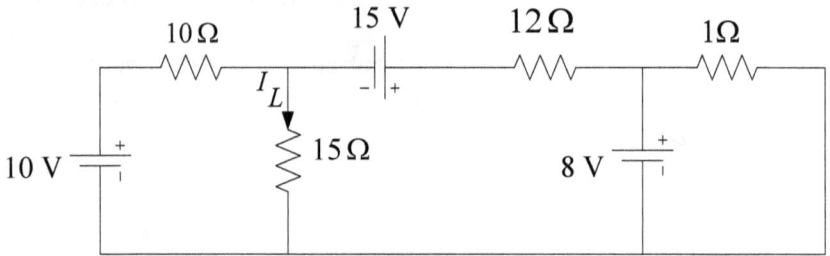

Fig. 2. Circuit diagram

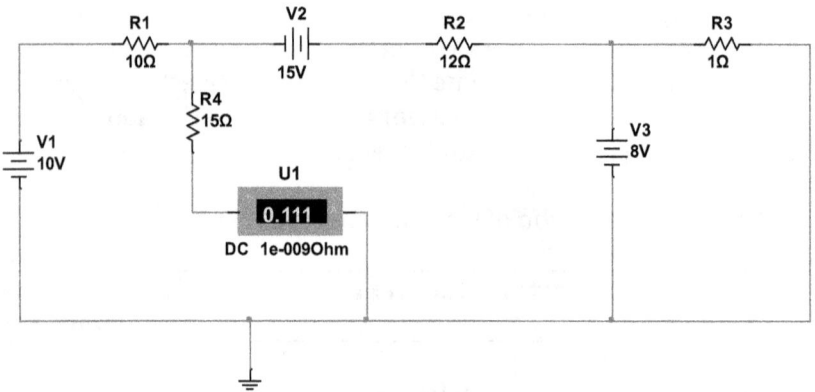

Fig. 3. Circuit connection in MULTISIM

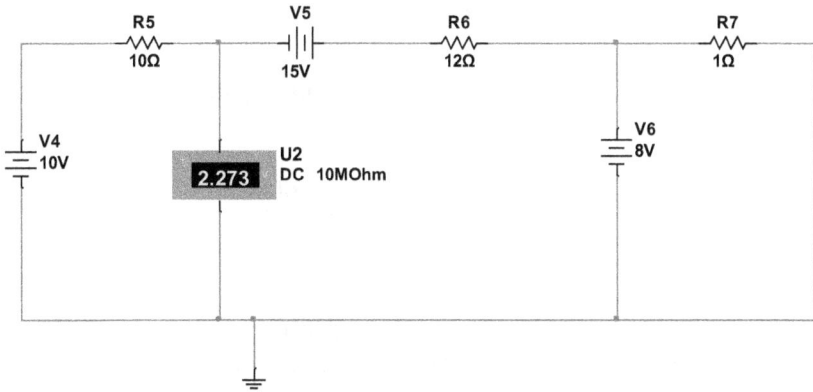

Fig. 4. Measurement of open circuit voltage

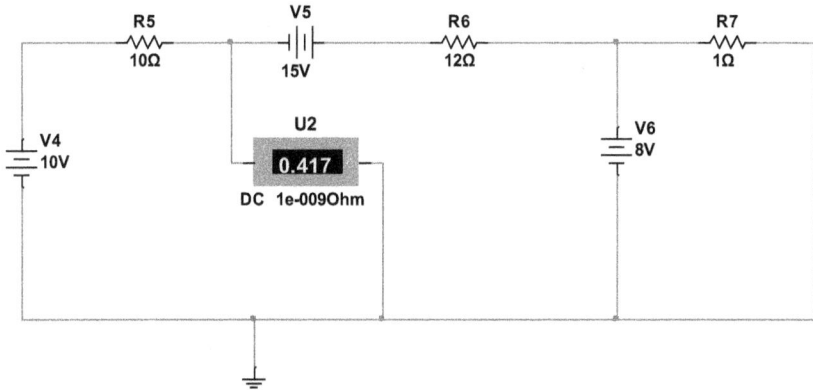

Fig. 5. Measurement of short circuit current

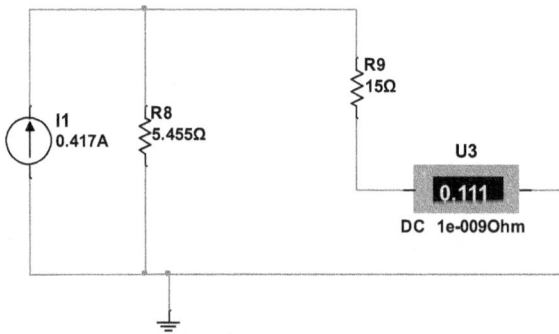

Fig. 6. Current measurement using Norton's equivalent circuit

EXPERIMENT 4

4.1 Objective: To verify Maximum Power Transfer theorem of a network.

4.2 Software Used: MULTISIM or similar software.

4.3 Brief Theory:

Maximum power transfer theorem is used to find the value of load impedance for which power transferred is maximum from source to connected load. For the DC circuit, statement of Maximum power transfer theorem is:

Power transferred from the source to load is maximum, if load resistance is equal to the internal resistance (i.e. Thevenin's equivalent resistance) of the source network seen from the load terminals.

Considering load voltage V_L and current I_L, power transferred to load resistance is given as

$$P_o = I_L^2 R_L \tag{4.1}$$

$$= \left(\frac{V_{Th}}{R_{Th}+R_L}\right)^2 R_L = \frac{V_{Th}^2 R_L}{R_{Th}^2 + 2R_{Th}R_L + R_L^2} = \frac{V_{Th}^2}{(R_{Th}^2/R_L) + 2R_{Th} + R_L} \tag{4.2}$$

Output power P_o will be maximum, when denominator (x) in equation (4.2) is minimum.

$$x = \left(R_{Th}^2/R_L\right) + 2R_{Th} + R_L \tag{4.3}$$

Hence, for minimum value of denominator (x)

$$\frac{dx}{dR_L} = 0 \text{ and } \frac{d^2x}{dR_L^2} \text{ is positive}$$

$$\frac{d}{dR_L}\left(\left(R_{Th}^2/R_L\right) + 2R_{Th} + R_L\right) = 0$$

$$\text{or } -\frac{R_{Th}^2}{R_L^2} + 0 + 1 = 0$$

or $\dfrac{R_{Th}^2}{R_L^2} = 1$

Therefore,

$$R_L = R_{Th} \qquad\qquad\qquad (4.4)$$

Further, second derivative of denominator x (as given in equation 4.5) is positive, showing condition given in equation 4.4 is valid for minimum value of x, and hence for maximum output power,

$$\dfrac{d^2x}{dR_L^2} = \dfrac{2}{R_{Th}} \qquad\qquad\qquad (4.5)$$

Hence, power transferred to load is maximum if $R_L = R_{Th}$.

4.4 Experiment Procedure using MULTISIM

Step 1: Open MULTISIM and create a new design file to develop the circuit.

Step 2: Make the connections as given in the circuit diagram of Fig. 2 by using MULTISIM. Circuit connection using MULTISIM is shown in Fig. 3 (a).

Step 3: Vary the load resistance and note the readings of load voltage V_L and current I_L, and calculate the output power P_{o1}. Power is also measured by using a wattmeter as shown in Fig. 3 (b). Error between the calculated and measured value of output power is determined by using equation 6. Readings at different value of load resistance is noted down in observation Table I.

$$E = \dfrac{P_{o1} - P_{o2}}{P_{o1}} \times 100 \qquad\qquad\qquad (4.6)$$

Step 4: Plot the measured output power at different value of load resistance as shown in Fig. 4. Graphically, output power is maximum, when load resistance is equal to 5.455 Ω (i.e. Thevenin's equivalent resistance).

4.5 Results:

Maximum power transfer theorem of the given circuit is verified using MULTISIM.

Readers may vary one or more voltage source and determine the maximum power transferred to load resistance. In all the cases, power transfer will be maximum, when $R_L = R_{Th}$. Further, it is suggested to develop MULTISIM model for other networks and follow the above procedural steps to verify the Maximum power transfer theorem.

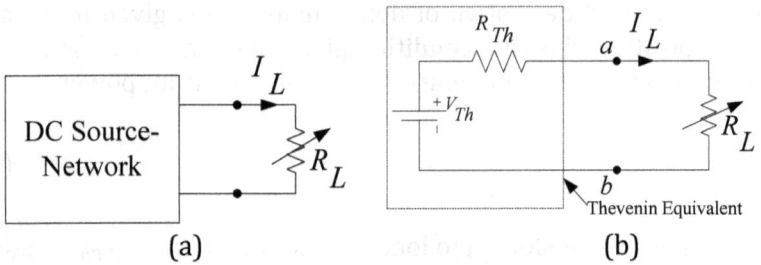

Fig. 1. (a) Original circuit (b) Equivalent source network and load

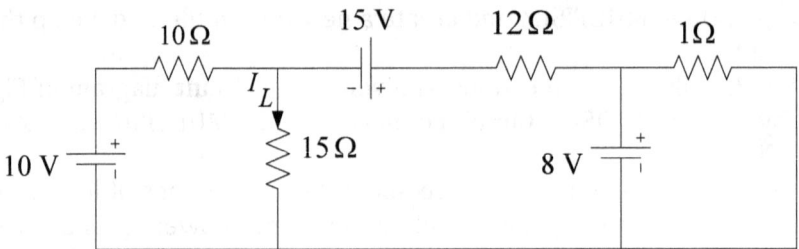

Fig. 2. Circuit diagram

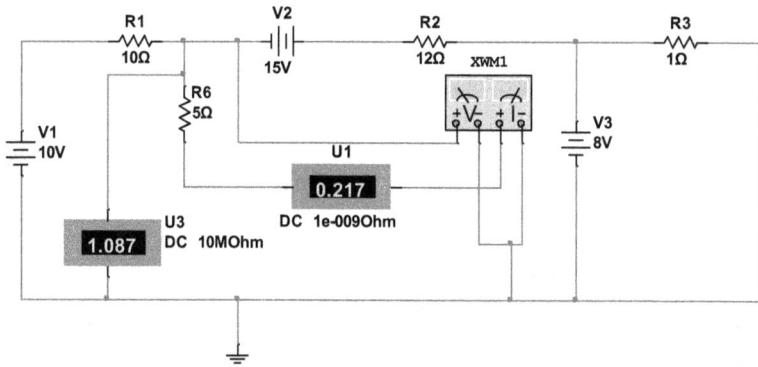

(a)

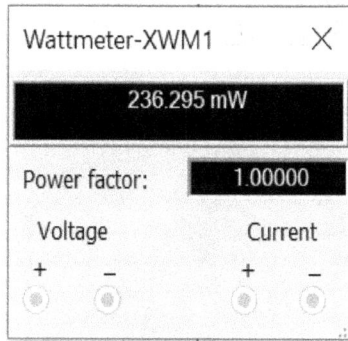

(b)

Fig. 3 (a). Circuit connection in MULTISIM (b) Wattmeter reading

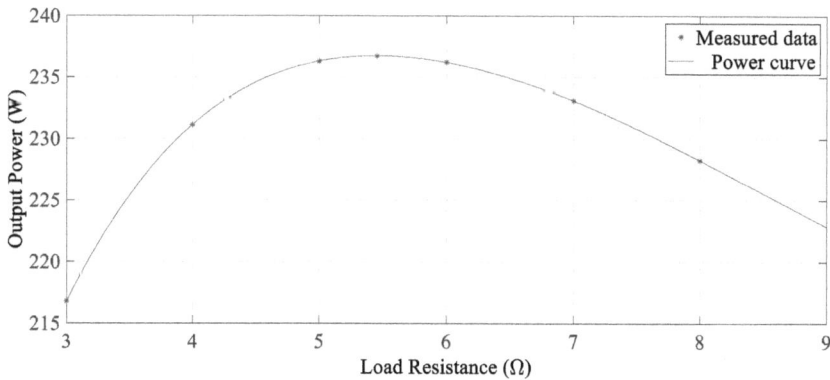

Fig. 4. Plot of output power Vs load resistance

Table I: Observation of Maximum power transfer theorem

S. No	Load resistance R_L (Ω)	Output voltage V_L (V)	Load current I_L (A)	Calculated output power P_{o1} (mW)	Measured output power P_{o2} (mW)	Percentage error E (%)
1	3	0.806	0.269	216.814	216.788	0.012
2	4	0.962	0.240	230.880	231.139	-0.112
3	5	1.087	0.217	235.879	236.295	-0.365
4	5.455	1.136	0.208	236.288	236.742	-0.058
5	6	1.190	0.198	235.620	236.205	-0.248
6	7	1.277	0.182	232.414	233.097	-0.294
7	8	1.351	0.169	228.319	228.269	0.022

EXPERIMENT 5

5.1 Objective: To verify of Tellegen's theorem of a network.

5.2 Software Used: MULTISIM or similar software.

5.3 Brief Theory

Tellegen's theorem: It states that the algebraic sum of powers in all branches in a network at any instant is zero. This theorem is valid for any network that may be linear or non-linear, active or passive and time varying or time invariant, and all branch currents and voltages in the network must satisfy Kirchhoff's laws. According to Tellegen's theorem, the rate of supply of energy by the active elements of a network equals the rate of energy dissipated or stored by the passive elements of the network. Tellegen's theorem is stated mathematically for b number of branches of network is as follows:

$$\sum_{K=1}^{b} V_K I_K = 0 \qquad\qquad (5.1)$$

V_K and I_K should satisfy KVL and KCL, respectively. In the above expression b indicates the number of branches.

5.4 Experiment Procedure using MULTISIM

Step 1: Open MULTISIM and create a new design file to develop the circuit.
Step 2: Make the connections as shown in the circuit diagram in Fig. 1 by using MULTISIM. Circuit connection using MULTISIM is shown in Fig. 2.
Step 3: Assuming the clockwise direction of current flow in each mesh, measure the value of current and voltage in each branch, and note down the readings in observation Table I.
Step 4: Verify KVL in each mesh.
Step 5: Verify KCL at each node.
Step 6: Verify Tellegen's theorem of the circuit.

5.5 Sample Calculation

5.5.1 Verification of KVL:

In mesh ABCFA:
$$V_{AB} + V_{BC} + V_{CF} + V_{FA}$$
$$= 12 - 4.696 - 7.304$$
$$= 0$$
In mesh CDEFC:
$$V_{CD} + V_{DE} + V_{EF} + V_{FC}$$
$$= -3.130 - 4.174 + 0 + 7.304$$
$$= 0$$

5.5.2 Verification of KCL:

At node C
$$I_{BC} + I_{CD} + I_{CF} = 4.696 - 3.652 - 1.043$$
$$= 0$$

1.5.3   Verification of Tellegen's theorem:
$$V_{AB}I_{AB} + V_{BC}I_{BC} + V_{CD}I_{CD} + V_{DE}I_{DE} + V_{CF}I_{CF}$$
$$= 12 \times 4.696 + (-4.696 \times 4.696) + (-3.13 \times 1.043)$$
$$+(-4.174 \times 1.043) + (-7.304 \times 3.652)$$
$$= 0$$

5.6 Results:

Tellegen's theorem of the given circuit is verified using MULTISIM.

Readers are suggested to develop MULTISIM model for other networks and follow the above procedural steps to verify the Tellegen's theorem. Readings may be noted in the format shown in Table II.

Table I: Branch voltage and current readings

S. No.	Branch	Voltage (V)	Current (A)	V*I
1.	AB	12.00	4.696	56.352
2.	BC	-4.696	4.696	-22.052
3.	CD	-3.130	1.043	-3.265
4.	CF	-7.304	3.652	-26.674
5.	DE	-4.174	1.043	-4.353

Table II: Branch voltage and current readings

S. No.	Branch	Voltage (V)	Current (A)	V*I

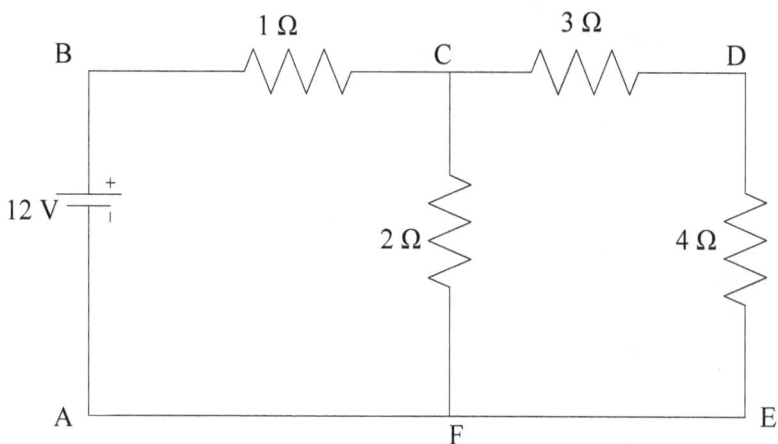

Fig. 1. Circuit diagram

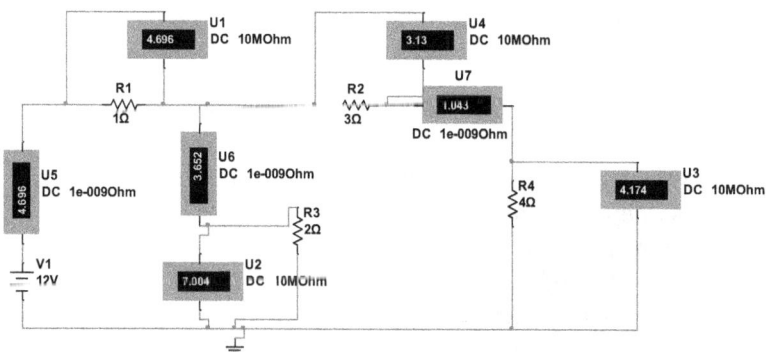

Fig. 2. Circuit connection in MULTISIM

EXPERIMENT 6

6.1 Objective: To determine Z and Y-parameters of a two port network.

6.2 Software Used: MULTISIM or similar software.

6.3 Brief Theory

A general two port network is shown in Fig. 1. Currents I_1 and I_2 are assumed to flow into the network at terminal voltages V_1 and V_2 from the input port $1 - 1'$ and output port $2 - 2'$ respectively. *Open-circuit impedance parameters or Z-parameters* of two-port network are calculated by considering V_1 and V_2 as dependent variables with I_1 and I_2 as independent variables. Similarly, I_1 and I_2 can be taken as dependent variables and V_1 and V_2 are taken as independent variables to determine the *short-circuit admittance or Y-parameters*.

6.3.1 Z-parameters

Considering Z_{11}, Z_{12}, Z_{21} and Z_{22} are the Z-parameters. Terminal voltage V_1 and V_2 in terms of current I_1 and I_2 are expressed as follows:

$$V_1 = Z_{11}I_1 + Z_{12}I_2 \tag{6.1}$$

$$V_2 = Z_{21}I_1 + Z_{22}I_2 \tag{6.2}$$

where
Z_{11} is known as the driving-point impedance at the input port, when output port is open circuited.
Z_{21} is known as the transfer impedance at the input port, when output port is open-circuited.
Z_{22} is known as the driving-point impedance at the output port, when input port is open circuited.
Z_{12} is known as the transfer impedance at the output port, when input port is open-circuited.

6.3.2 Y-parameters

Considering Y_{11}, Y_{12}, Y_{21} and Y_{22} are the Y-parameters. Terminal current I_1 and I_2 in terms of voltage V_1 and V_2 are expressed as follows:

$$I_1 = Y_{11}V_1 + Y_{12}V_2 \qquad\qquad (6.3)$$

$$I_2 = Y_{21}V_1 + Y_{22}V_2 \qquad\qquad (6.4)$$

where
Y_{11} is also known as the driving-point admittance at input port with output port short-circuited.
Y_{21} is also known as the transfer admittance at input port with output port short-circuited.
Y_{12} is also known as the transfer admittance at output port with input port short-circuited.
Y_{22} is also known as the driving-point admittance at output port with input port short-circuited.

6.3 Experiment Procedure using MULTISIM

Step 1: Open MULTISIM and create a new design file to develop the circuit.
Step 2: Make the connections as shown in the circuit diagram in Fig. 2 by using MULTISIM.
Step 3: First open the output terminals $2 - 2'$ and supply 12 V to input terminals $1 - 1'$. Measure output voltage and input current in developed MULTISIM model, as shown in Fig. 2 (a).
Step 4: Secondly, open the input terminals $1 - 1'$ and supply 12 V to output terminals $2 - 2'$, as shown in Fig. 2 (b). Measure input voltage and output current. Readings for Z-parameters are noted in observation Table I.
Step 5: Calculate the values of Z parameters using equation (6.1) and (6.2).
Step 6: Then short circuit the output terminal $(2 - 2')$ and supply 12 V to input terminal $(1 - 1')$. Measure input and output current.
Step 7: After that, short circuit the input terminal $(1 - 1')$ and supply 12V to output terminal $(2 - 2')$. Measure input and output current. Measure input voltage and output current. Readings for Y-parameters are noted in observation Table II.

123

Step 8: Calculate the values of Y parameters using equation (6.3) and (6.4).

6.5 Sample Calculation:

6.5.1 For Z Parameters

(1)   When output terminals (2–2') are opened circuited i.e. $I_2 = 0$

$$Z_{11} = \frac{V_1}{I_1} = \frac{12.0}{3.75} = 3.2\Omega$$

$$Z_{21} = \frac{V_2}{I_1} = \frac{9.0}{3.75} = 2.4\Omega$$

(2) When input terminals (1–1') are opened circuited i.e. $I_1 = 0$

$$Z_{12} = \frac{V_1}{I_2} = \frac{6.0}{2.5} = 2.4\Omega$$

$$Z_{22} = \frac{V_2}{I_2} = \frac{12.0}{2.5} = 4.8\Omega$$

6.5.2 For Y parameters

(1) When output terminals (2–2') are short circuited i.e. $V_2 = 0$

$$Y_{11} = \frac{I_1}{V_1} = \frac{6.0}{12.0} = 0.5\ \mho$$

$$Y_{21} = \frac{I_2}{V_1} = \frac{-3.0}{12.0} = -0.25\ \mho$$

(2) When input terminals (1–1') are short circuited i.e. $V_1 = 0$

$$Y_{12} = \frac{I_1}{V_2} = \frac{-3.0}{12.0} = -0.25\ \mho$$

$$Y_{22} = \frac{I_2}{V_2} = \frac{4.0}{12.0} = 0.33\ \mho$$

6.6 Results:

The Z and Y-parameters of a two port network has been calculated.

Calculated parameters (both Z and Y) are based on the experimental readings using software. Readers are suggested to verify the parametric values by theoretical calculation. It is also suggested to perform the experiment with other two-port network and note down the readings in Table III and IV.

Table I: Readings for Z-parameters

S. No.	Output (2−2′) opened circuited		
	V_2	V_1	I_1
1	9.0	12.0	3.75
	Input (1−1′) opened circuited		
	V_2	V_1	I_2
2	12.0	6.0	2.5

Table II: Readings for Y-parameters

S. No.	Output (2 − 2′) short circuited		
	V_1	I_1	I_2
1	12.0	6.0	-3.0
	Input (1−1′) short circuited		
	V_2	I_1	I_2
2	12.0	-3.0	4.0

Table III: Readings for Z-parameters

S. No.	Output (2−2′) opened circuited		
	V_2	V_1	I_1
	Input (1−1′) opened circuited		
	V_2	V_1	I_2

Table IV: Readings for Y-parameters

S. No.	Output (2 − 2′) short circuited		
	V_1	I_1	I_2
	Input (1−1′) short circuited		
	V_2	I_1	I_2

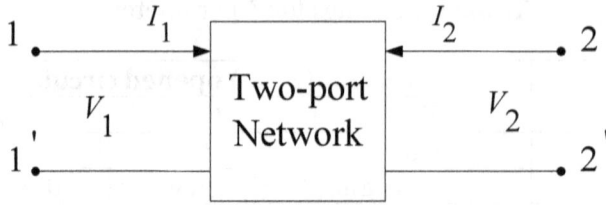

Fig. 1. Two-port network

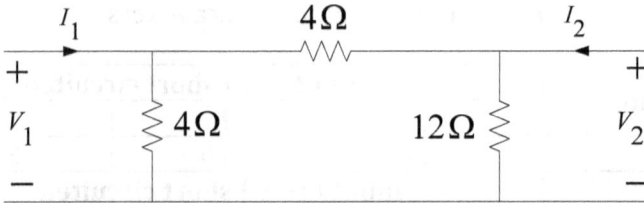

Fig. 2. Circuit for determining two-port parameters

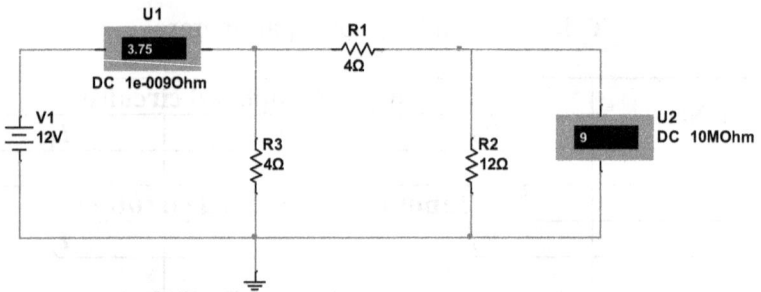

(a)

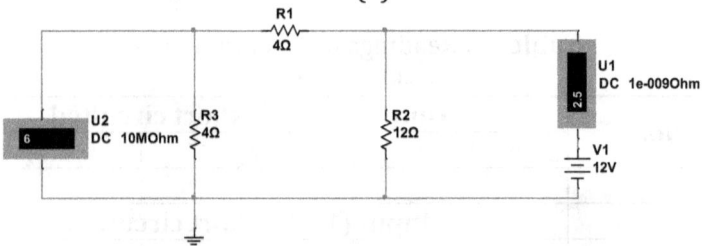

(b)

Fig. 3. Two port network in MULTISIM (a) output terminals opened (b) input terminals opened

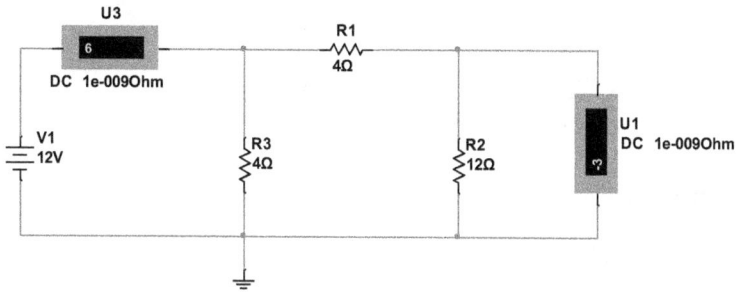

(a)

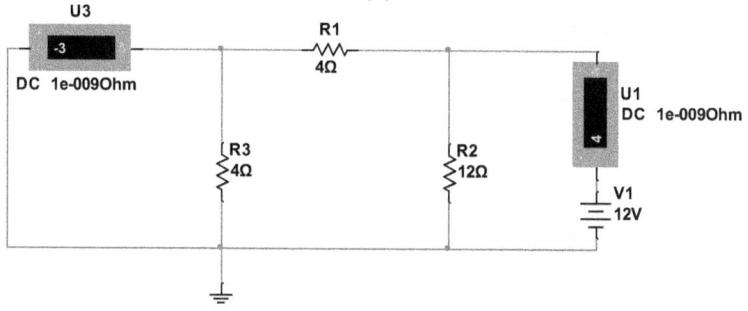

(b)

Fig. 4. Two port network in MULTISIM (a) output terminals shorted
(b) input terminals shorted

127

EXPERIMENT 7

7.1 Objective: To determine transient response of current in RL circuits with step voltage input.

7.2 Software Used: MULTISIM or similar software.

7.3 Brief Theory:

In electrical circuit, resistor, inductor and capacitor are considered as basic elements for circuit design. Unlike to the resistor, current do not increases instantaneously in capacitor and inductor, which are an energy storage element. In a series RL circuit, current increases smoothly with a time constant ($\tau = {}^{L}/_{R}$) given by

$$I = I_0\left(1 - e^{-t/\tau}\right) \tag{7.1}$$

where

$$I_0 = \frac{V}{R} \tag{7.2}$$

Hence, time constant τ may be expressed as

$$\tau = \frac{t}{\log\left(\frac{I_0}{I_0 - I}\right)} \tag{7.3}$$

In this experiment, circuit current flowing through RL circuit is noted at different time to evaluate the time constant τ and response is plotted.

7.4 Experiment Procedure using MULTISIM

Step 1: Open MULTISIM and create a new design file to develop the circuit.
Step 2: Make the connections as shown in circuit diagram given in Fig. 1 by using MULTISIM. Circuit connection using MULTISIM is shown in Fig. 2.
Step 3: Use the suitable setting of connected oscilloscope and probe. Such setting is shown with current response in Fig. 3.

Step 4: Note down the value of current at different time instant as shown in observation Table-I.

Step 5: Calculate the value of time constant at each recorded reading as shown in section 7.5. Take the average value of calculated time constant at each recorded reading.

7.5 Sample calculation

With input voltage at 50 V in circuit shown in Fig. 1,
$$I_0 = \frac{50}{10} = 5\ A\ .$$
At time $t = 0.056$ sec.,
$$\tau = \frac{t}{\log\left(\frac{I_0}{I_0 - I}\right)} = \frac{0.056}{\log\left(\frac{5}{5 - 2.134}\right)} = 0.10$$

7.6 Results:

1) Step response of RL circuit was obtained using MULTISIM.

2) Time constant of given RL circuit was determined experimentally using MULTISIM.

Readers are suggested to follow the above procedure with different parameters values and other circuit with RL elements to obtain transient response.

Table I: Observed current at different time instant

S. No.	Current (A)	Time (Sec.)	Time constant τ
1.	2.13	0.06	0.10
2.	3.65	0.13	0.10
3.	4.36	0.21	0.10
4.	4.71	0.28	0.10

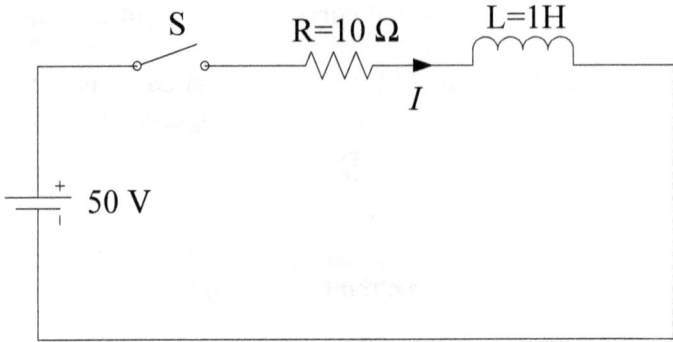

Fig. 1. Circuit diagram

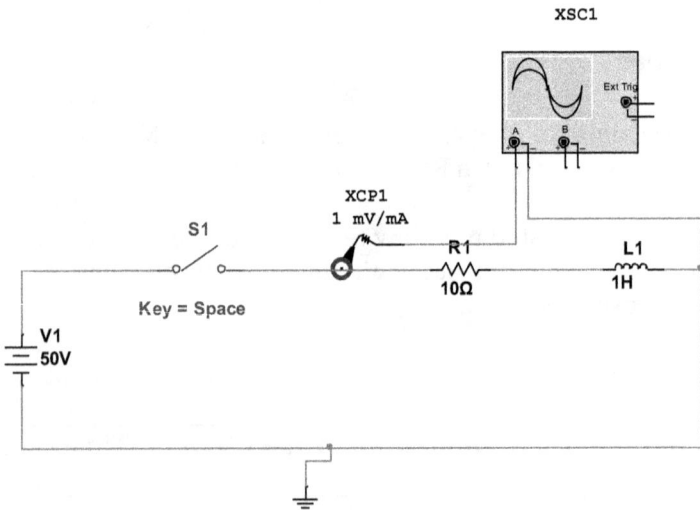

Fig. 2. Circuit connection in MULTISIM

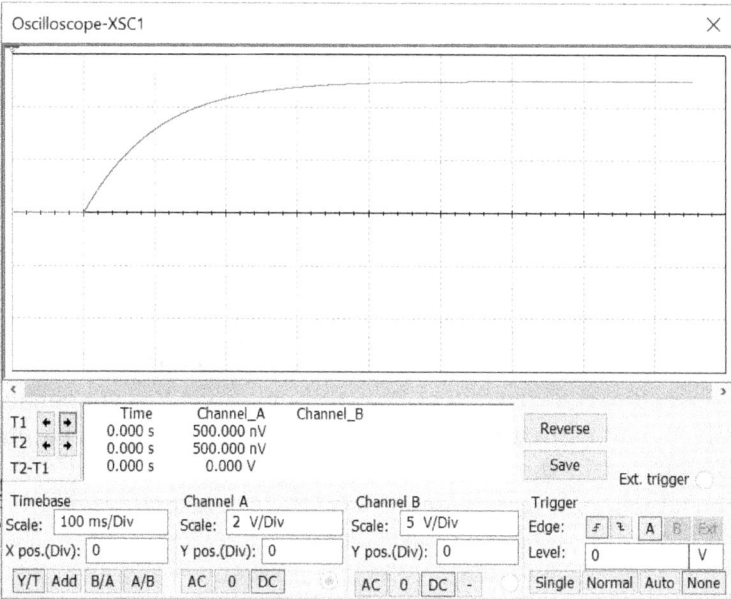

Fig. 3. Oscilloscope setting

EXPERIMENT 8

8.1 Objective: To determine transient response of current in RC circuits with step voltage input.

8.2 Software Used: MULTISIM or similar software.

8.3 Brief Theory:

Consider a simple series RC circuit connected to a DC source though a single pole single through switch (S) as shown in Fig. 1. Initially, after operating the switch S to ON position, capacitor will behave as a short circuit with their both parallel plates at same potential. Hence, current in the circuit will be maximum (i.e. $I_0 = V/R$). Capacitor will start to get charged with charge accumulation on both parallel plates. Hence, voltage across capacitor will increases smoothly with respect to time which is mathematically given by

$$v_c = V\left(1 - e^{\frac{-t}{\tau}}\right) \tag{8.1}$$

where

V is the input voltage and τ is the time constant ($\tau = RC$).

Hence, time constant τ may be expressed as

$$\tau = \frac{t}{\log\left(\frac{V}{V-v_c}\right)} \tag{8.2}$$

In this experiment, capacitor voltage build-up across the RC circuit is noted at different time to evaluate the time constant τ and to plot the response.

8.4 Experiment Procedure using MULTISIM

Step 1: Open MULTISIM and create a new design file to develop the circuit.
Step 2: Make the connections as shown in circuit diagram depicted in Fig. 1 by using MULTISIM. Circuit connection using MULTISIM is shown in Fig. 2.

Step 3: Take the circuit parameters as: $R = 1k\Omega, C = 100\mu F$. Use the suitable setting of connected oscilloscope and probe. Such setting is shown with current response in Fig. 3.

Step 4: Note down the value of capacitor voltage at different time instant as shown in observation Table-I.

Step 5: Calculate the value of time constant at each recorded reading as shown in section 8.3. Take the average value of calculated time constant at each recorded reading. This value was found to be 0.096 sec. for the considered circuit, which is in closed agreement with theoretical value ($\tau = RC = 0.1sec.$).

8.5 Sample calculation

With input voltage at 10 V in circuit shown in Fig. 1,
Input voltage $V = 10$ volt
At time $t = 0.074$ sec.

$$\tau = \frac{t}{\log\left(\frac{V}{V-v_c}\right)} = \frac{0.074}{\log\left(\frac{10}{10-5.597}\right)} = 0.0902 \text{ sec.}$$

8.6 Conclusions:

1) Step response of RC circuit was obtained using MULTISIM.
2) Time constant of given RC circuit was determined experimentally using MULTISIM.

Readers are suggested to follow the above procedure with different parameters values and other circuit with RC elements to obtain transient response.

Table I: Observed capacitor voltage at different time instant

S. No.	Voltage (v_c)	Time (Sec.)	Time constant (τ)
1.	5.597	0.074	0.090
2.	6.570	0.099	0.093
3.	8.418	0.176	0.100
4.	0.326	9.646	0.100

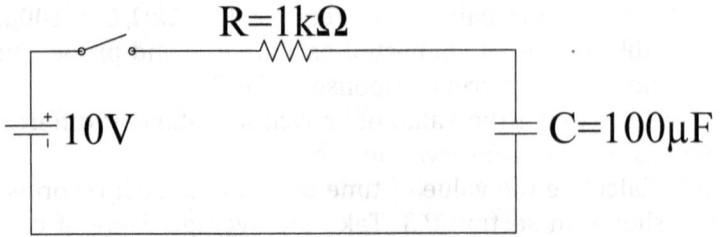

Fig.1. Circuit diagram

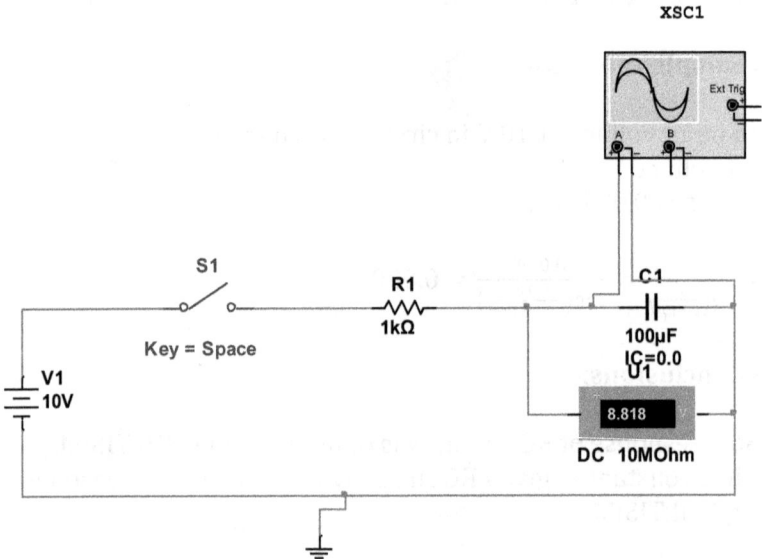

Fig. 2. Circuit connection in Multisim

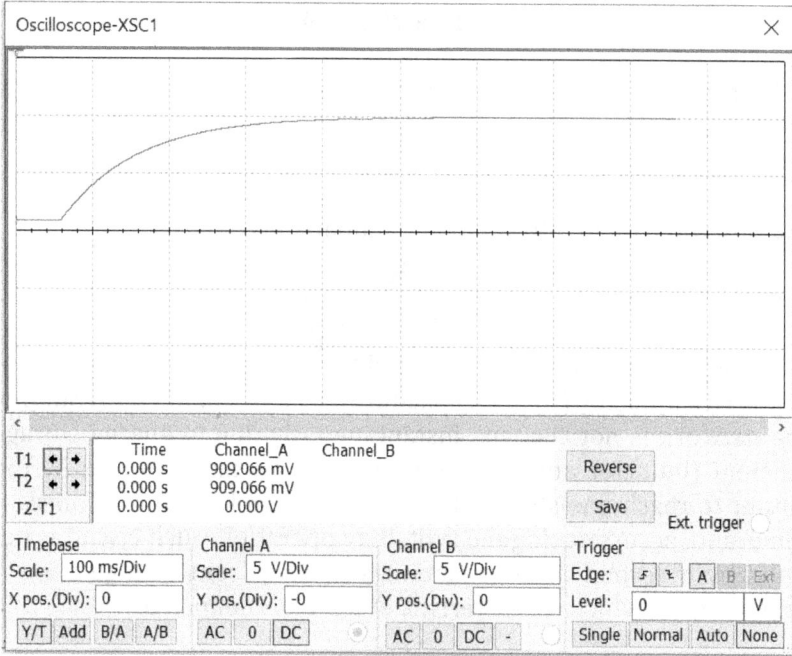

Fig. 3. Circuit response in oscilloscope

EXPERIMENT 9

9.1 Objective: To determine transient response of current in RLC circuit with step voltage input for underdamped, critically damped and overdamped cases.

9.2 Software Used: MULTISIM or similar software.

9.3 Brief Theory:

A simple RLC network is constituted by the series connection of resistor (R), inductor (L) and capacitor (C) as shown in Fig. 1. State of the network is not changed instantaneously due to energy storage element (inductor and capacitor). Hence, the time is taken by the circuit to reach one steady state to another is known as *'Transient Time'* and, occurs during the switching operation. Such circuit is defined by a second order equation, obtained from voltage-current relation from which characteristic equation is defined as

$$s^2 + \frac{R}{L}s + \frac{1}{LC} = 0 \qquad\qquad (9.1)$$

and their roots are given as

$$s = -\frac{R}{2L} \pm \sqrt{\left(\frac{R}{2L}\right)^2 - \frac{1}{LC}} \qquad\qquad (9.2)$$

Hence, the circuit response depends on following cases:

Case I: $\left(\frac{R}{2L}\right)^2 > \frac{1}{LC}$ The roots are real and unequal and it gives an overdamped response.

Case II: $\left(\frac{R}{2L}\right)^2 < \frac{1}{LC}$ The roots are complex conjugate and it gives an underdamped response.

Case III: $\left(\frac{R}{2L}\right)^2 = \frac{1}{LC}$ The roots are real and equal and it gives a critically damped response.

9.4 Experiment Procedure using MULTISIM

Step 1: Open MULTISIM and create a new design file to develop the circuit.

Step 2: Make the connections as shown in the circuit diagram in Fig. 1 by using MULTISIM. Circuit connection using MULTISIM is shown in Fig. 2.

Step 3: Take the circuit parameters as: $L = 1H$, $C = 0.5F$ and step input voltage at 12 V. For overdamped case, resistance $R = 5\Omega$ is used. Use the suitable setting of connected oscilloscope as shown in Fig. 3. Current response of the circuit is shown in Fig. 3 (a). Terminal voltage across inductor and capacitor is shown in Fig. 3 (b). Inductor voltage was decreased to zero as shown by red curve and capacitor was in charging mode which increases to input source as shown by black curve.

Step 4: For underdamped case, resistance $R = 1\Omega$ is used. Use the suitable setting of connected oscilloscope as shown in Fig. 4. Current response of the circuit is shown in Fig. 4 (a). Terminal voltage across inductor and capacitor is shown in Fig. 4 (b). Inductor voltage was decreased to zero as shown by red curve and capacitor was in charging mode which increases to input source as shown by black curve. It may be noted that the settling time of circuit response is more than the overdamped case of step 3.

Step 5: For critically damped case, resistance $R = 2.83\Omega$ is used. Use the suitable setting of connected oscilloscope as shown in Fig. 5. Current response of the circuit is shown in Fig. 5 (a). Terminal voltage across inductor and capacitor is shown in Fig. 5 (b). Inductor voltage was decreased to zero as shown by red curve and capacitor was in charging mode which increases to input source as shown by black curve. Comparatively, peak overshoot of circuit current was found to be maximum for critically damped case.

9.5 Result:

Step response of RLC circuit was obtained and plotted using MULTISIM.

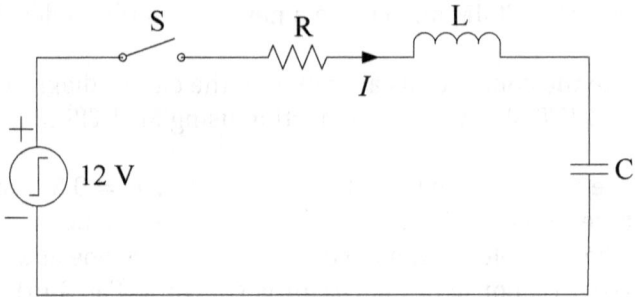

Fig. 1. Circuit diagram

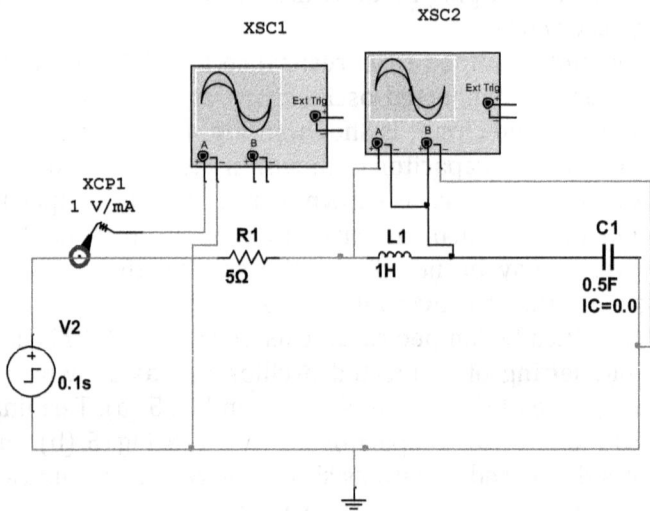

Fig. 2. Circuit connection in MULTISIM

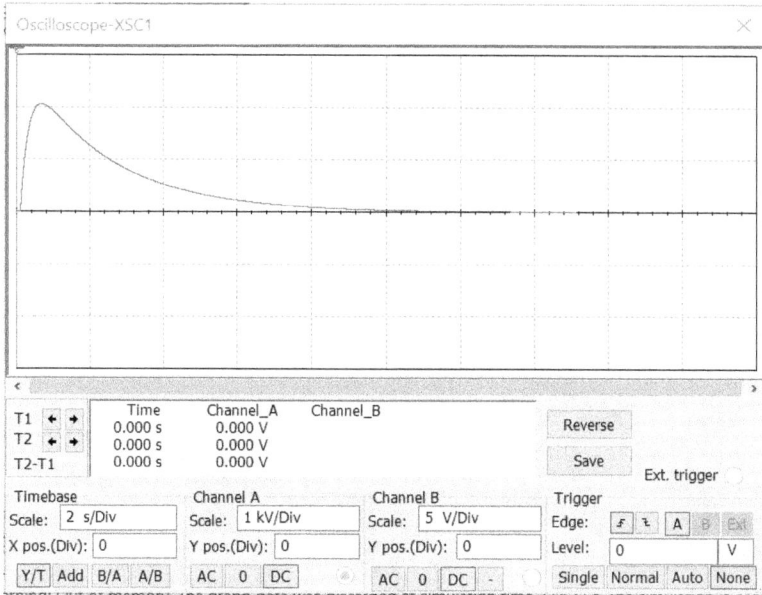

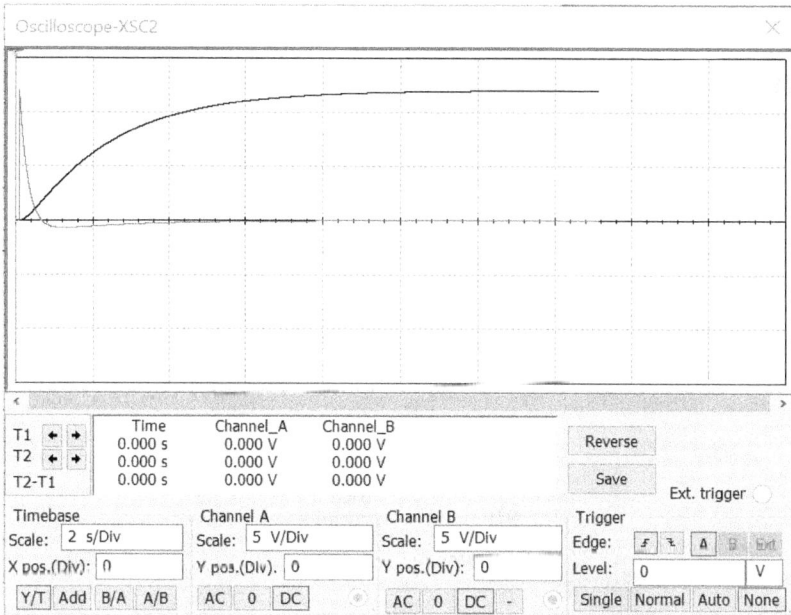

Fig. 3. Circuit response in overdamped condition (a) circuit current (b) voltage across inductor and capacitor

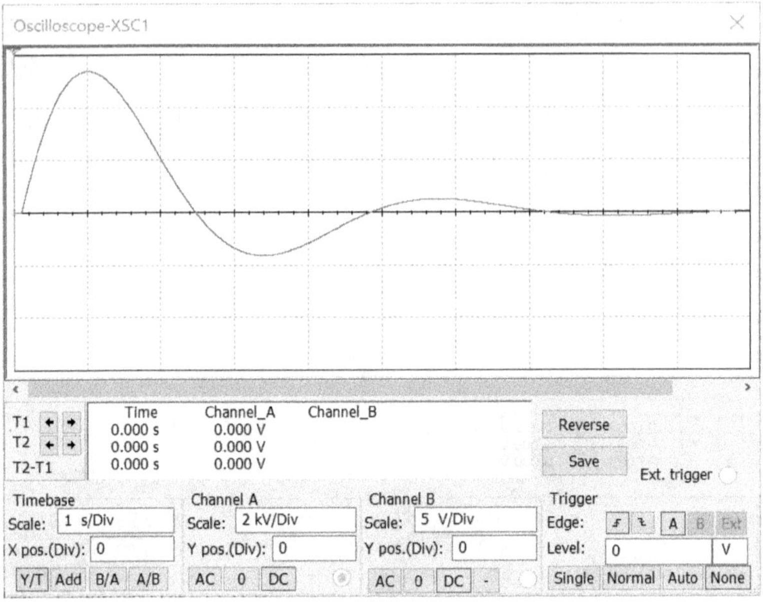

Fig. 4. Circuit response in underdamped condition (a) circuit current (b) voltage across inductor and capacitor

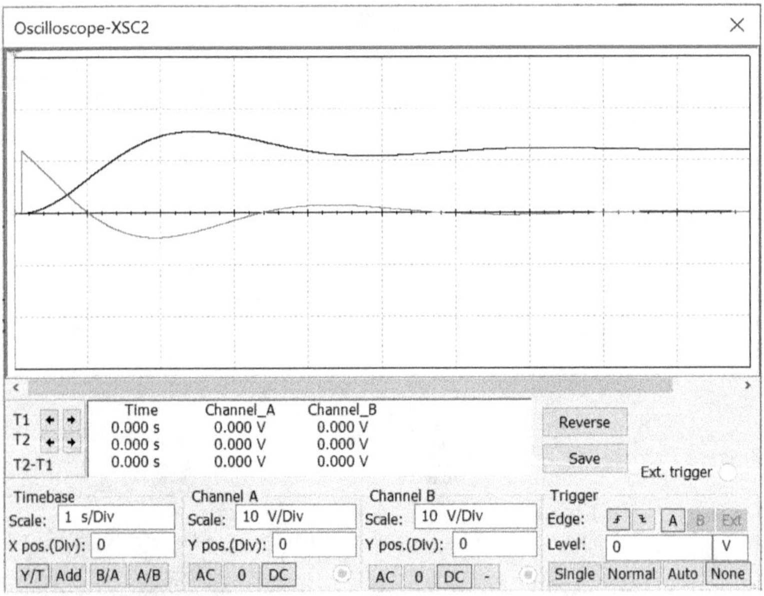

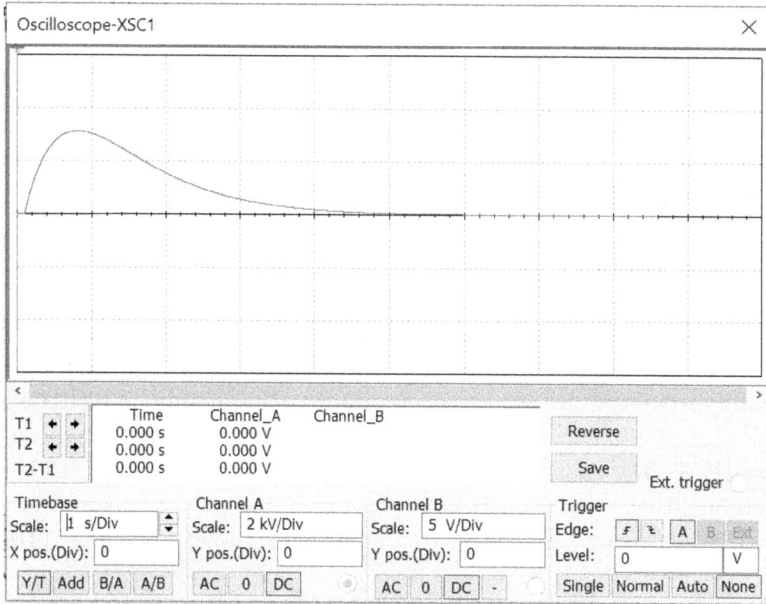

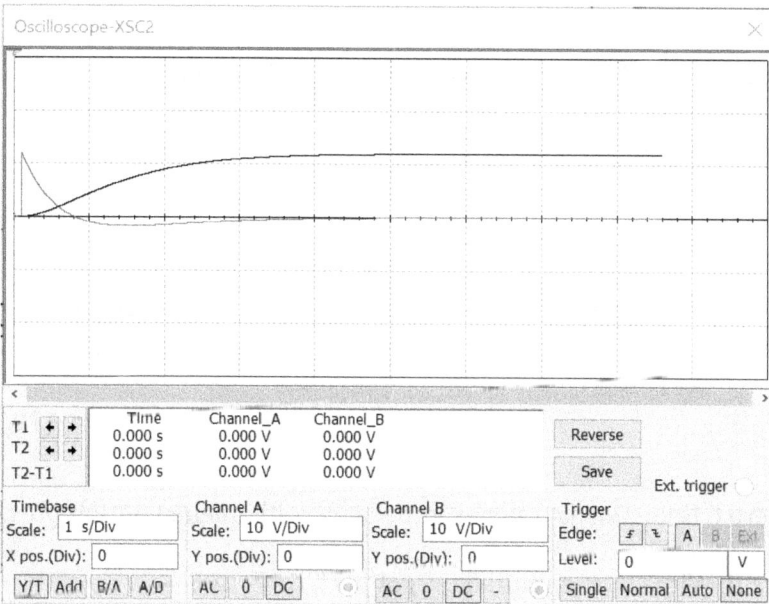

Fig. 5. Circuit response in critically-damped condition (a) circuit current (b) voltage across inductor and capacitor

141

EXPERIMENT 10

10.1 Objective: To determine transmission or ABCD-parameters of a two port network.

10.2 Software Used: MULTISIM or similar software.

10.3 Brief Theory:

In a two-port network, if the direction of output terminal is assumed to reversed (i.e. directed from the network) as shown in Fig. 1, voltage-current at input as dependent variable may be written as

$$V_1 = AV_2 - BI_2 \tag{10.1}$$

$$I_1 = CV_2 - DI_2 \tag{10.2}$$

Above expressions may be expressed in matrix form as follows:

$$\begin{bmatrix} V_1 \\ I_1 \end{bmatrix} = \begin{bmatrix} A & B \\ C & D \end{bmatrix} \begin{bmatrix} V_2 \\ -I_2 \end{bmatrix} \tag{10.3}$$

ABCD/Transmission parameters are defined as follows:

Case I: With open-circuited output port, i.e. $I_2 = 0$

$$A = \frac{V_1}{V_2}\bigg|_{I_2=0} \tag{10.4}$$

where A is known as reversed voltage gain with opened output port.

$$C = \frac{I_1}{V_2}\bigg|_{I_2=0} \tag{10.5}$$

where C is known as transfer impedance with opened output port.

Case II: With short-circuited output port, i.e. $V_2 = 0$

$$B = -\frac{V_1}{I_2}\bigg|_{V_2=0} \tag{10.6}$$

where B is known as transfer function with shorted output port.

142

$$D = -\frac{I_1}{I_2}\Big|_{V_2=0} \tag{10.7}$$

where D is known as reversed current gain with shorted output port.

10.4 Experiment Procedure using MULTISIM

Step 1: Open MULTISIM and create a new design file to develop the circuit.
Step 2: Make the connections as shown in the circuit diagram in Fig. 2 by using MULTISIM.
Step 3: First open the output terminals $2 - 2'$ and supply 12 V to input terminals $1 - 1'$. Measure output voltage and input current in developed MULTISIM model, as shown in Fig. 3 (a).
Step 4: Then short circuit the output terminal $(2 - 2')$ and supply 12 V to input terminal $(1 - 1')$. Measure input and output current in developed MULTISIM model, as shown in Fig. 3 (b).
. Readings for ABCD-parameters are noted in observation Table I.
Step 5: Calculate the values of ABCD parameters using equation (10.4) to (10.7).

10.5 Sample Calculation:

Case I: With open-circuited output port, i.e. $I_2 = 0$
$$A = \frac{V_1}{V_2}\Big|_{I_2=0} = \frac{12}{10} = 1.2$$
$$C = \frac{I_1}{V_2}\Big|_{I_2=0} = \frac{2}{10} = 0.25$$
Case II: With short-circuited output port, i.e. $V_2 = 0$
$$B = -\frac{V_1}{I_2}\Big|_{V_2=0} = -\frac{12.0}{3.53} = -3.4$$
$$D = -\frac{I_1}{I_2}\Big|_{V_2=0} = -\frac{4.94}{3.53} = -1.4$$

10.6 Result:

The ABCD-parameters of the two port network has been calculated by using experimental readings.

Readers are suggested to verify the parametric values by theoretical calculation. It is also suggested to perform the experiment with other two-port network and note down the readings in Table II.

Table I: Readings for ABCD-parameters

S. No.	Output (2−2′) opened circuited		
	V_1	V_2	I_1
1	12.0	10.0	2.0
	Output (2−2′) short circuited		
	V_1	I_1	I_2
2	12.0	4.94	3.53

Table II: Readings for ABCD-parameters

S. No.	Output (2−2′) opened circuited		
	V_1	V_2	I_1
	Output (2−2′) short circuited		
	V_1	I_1	I_2

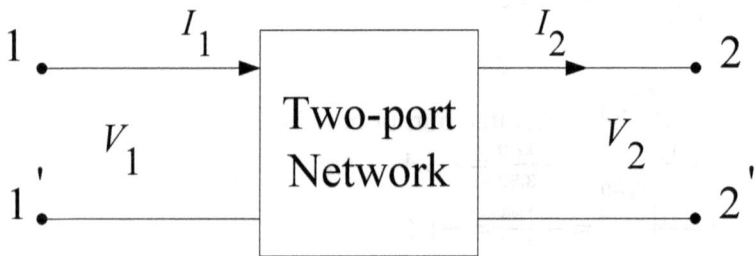

Fig. 1. Two-port network

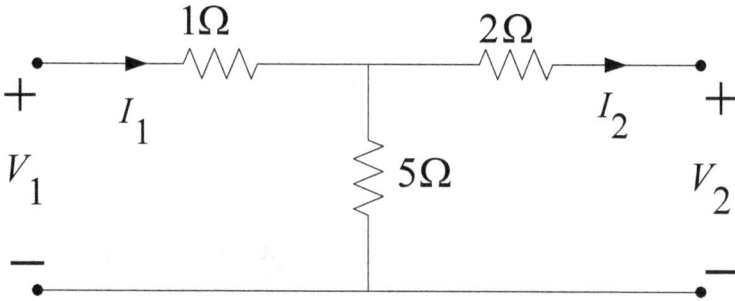

Fig. 2. Circuit for determining two-port parameters

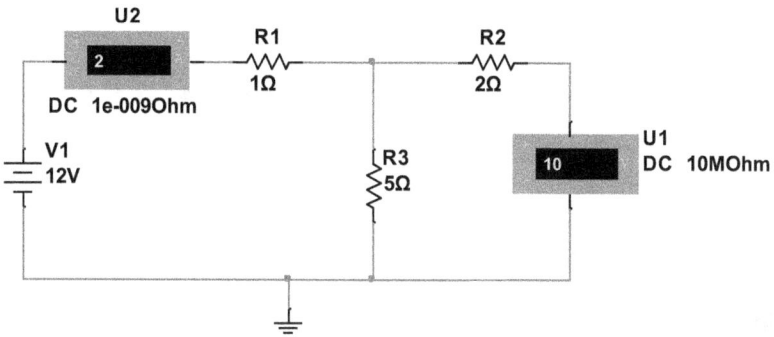

(a)

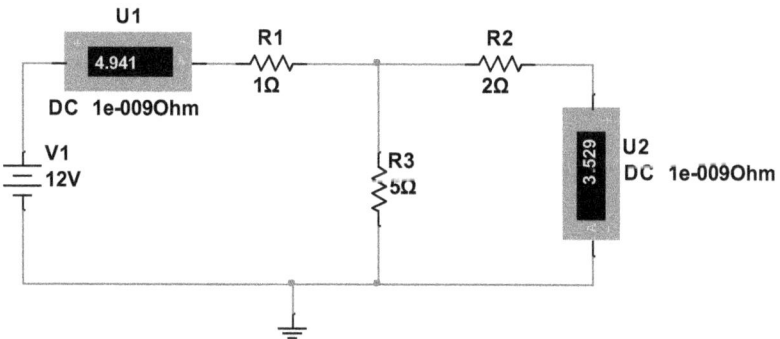

(b)

Fig. 3. Two port network in MULTISIM (a) output terminals opened
(b) output terminals shorted

References

1. S. K. Bhattacharya, M. Singh, "Network Analysis and Synthesis", Pearson Education India, 2017.
2. R. R. Singh, "Circuit Theory and Networks Analysis and Synthesis", McGraw-Hill Education, 2nd edition, 2018.
3. A. Husain, "Network and Systems", Khanna Book Publishing Co. (P) Ltd., New delhi, India, 2nd edition, 2019.
4. K. S. S. Kumar, "Electric Circuits and Networks", Pearson Education India, 2009.
5. NI Multisim, User Manual, National Instruments Corporation, USA, 2009.

About the Book

A Laboratory Manual on "**Virtual Experimentation on Electrical AC Machines and Circuit Networks using MATLAB/Simulink and MULTISIM**" has been written to quickly grip the understanding of Electrical AC machines, particularly induction and synchronous motor, and Circuit Networks from an experimental operation point of view on the virtual platform. This book deals with all the experiments related to important topics of electrical machines and networks which are practically performed by students in a majority of technical institutions. The theoretical background of every experiment is reviewed before the system simulation. Detailed step-by-step experimental procedures are explained with the necessary diagram. Development of experimental setup by using MATLAB/Simulink and MULTISIM has been explained from scratch, which also enhances the simulation and analytical skills for various practical systems. Every step of the simulation is explained during the development of the system using Simulink and MULTISIM which can be further extended to investigate and analyze for Academics, Industrial, and Research & Development purposes. This book can be used as a reference to simulate and analyze the virtually developed system using AC machines and circuit networks by students and working professionals.

ISBN 978-1-922617-41-5
90000

9 781922 617415

https://centralwestpublishing.com